Rethinking Economic Behaviour

Also by David Simpson

GENERAL EQUILIBRIUM ANALYSIS
THE POLITICAL ECONOMY OF GROWTH
THE END OF MACROECONOMICS?

Rethinking Economic Behaviour

How the Economy Really Works

David Simpson

First published in Great Britain 2000 by
MACMILLAN PRESS LTD
Houndmills, Basingstoke, Hampshire RG21 6XS and London
Companies and representatives throughout the world

A catalogue record for this book is available from the British Library.

ISBN 0–333–77926–6

 First published in the United States of America 2000 by
ST. MARTIN'S PRESS, LLC,
Scholarly and Reference Division,
175 Fifth Avenue, New York, N.Y. 10010

ISBN 0–333–77926–6

Library of Congress Cataloging-in-Publication Data
Simpson, David, 1936–
    Rethinking economic behaviour : how the economy really works / David
Simpson.
        p. cm.
    Includes bibliographical references and index.
    ISBN 0–333–77926–6
        1. Economics. 2. Capitalism. 3. System theory. I. Title.

    HB71 .S56 2000
    330.12'2—dc21
                                                                00–062699

This book is printed on paper suitable for recycling and made from fully managed and sustained forest sources.

10   9   8   7   6   5   4   3   2
09   08   07   06   05   04   03   02

Printed and bound in Great Britain by
Antony Rowe Ltd, Chippenham, Wiltshire

In Memoriam JWMcG

Contents

Preface

Many friends and colleagues have kindly read earlier drafts of chapters of this book. They include: Fessal Bouaziz, Arie De Geus, Paul Hare, Brian Loasby, Hugh Macmillan, Douglas Mair, John Moore, Sir Alan Peacock, Colin Robinson, Mark Schaffer, Chester Simpson and Paul Williamson. I am grateful to all of them for their helpful comments. They are of course absolved of any responsibility for the results. I should also like to express my thanks to colleagues in the Department of Economics at Heriot-Watt University for affording me the use of their facilities.

Some of the material in Chapter 15 first appeared in chapter 6 of a book edited by David Reisman under the title *Economic Thought and Political Theory*, published in 1994. I am most grateful to Kluwer Academic Publishers for granting me permission to use this material. I should like to thank my son, Donald, who brought his professional skills to bear at a critical moment in the production of the typescript. My deepest gratitude must go to my wife Barbara, not only for her skilful assistance in the preparation of this book, but above all for her loving support over many years.

DAVID SIMPSON

1
Introduction

'To explain economic activity in advanced human societies is like accounting for the course of a flowing stream which is constantly adapting itself to changes in circumstances of which each participant can know only a small fraction, and not for a hypothetical state of equilibrium determined by a set of ascertainable data.'

F.A Hayek, *Law, Legislation and Liberty.*

The purpose of this book is to put forward an alternative to neoclassical equilibrium theory. The limitations of equilibrium theory have become increasingly apparent in recent years, particularly to those economists who attempt to use it to explain the workings of an actual market economy, whether in offering policy advice to politicians and businessmen, or simply in interpreting events to a wider public. In response to rising dissatisfaction, neoclassical economists have offered some modest modifications to the standard theory, but these modifications have generally taken the form of accommodating new assumptions to the requirements of the existing model rather than to reality. This is not a satisfactory scientific procedure.

In the last ten years a quite different approach to economic theory, in which the market economy is treated as an evolving complex adaptive system, has emerged, but it has yet to capture the imagination of the profession or the wider public.[1] The rather cool reception which this approach has so far received may in

1

large part be explained by its presentation. It has been treated by its proponents as if it were a mathematical technique originating in the natural sciences which may have some limited applications in some specific aspects of economics and the other social sciences. In fact, as we shall show, its underlying principles of self-organisation and evolution can trace their intellectual origins to seventeenth and eighteenth-century philosophers of society. Its perspective is of perfectly general application to all branches of science. Furthermore, it does something which equilibrium theory, originating in nineteenth-century mechanics, fails to do, which is to distinguish clearly between human and non-human phenomena.

So far as economics is concerned, complexity theory (as we shall call it for short) provides a thoroughly unifying function. It is congruent with Austrian theories and the older classical tradition of Adam Smith. It thereby offers an approach to a unified social science. Complexity theory also provides a common mode of discourse with business studies as well as with the natural sciences.

Equilibrium theory offers the apparent possibility of prediction by achieving a spurious precision. It does not pretend to explain how a market economy actually works. Complexity theory, on the other hand, does attempt to explain how a market economy works, and in so doing it shows why precise prediction of particular economic events is impossible. Equilibrium theory rules out any direct interaction between economic agents: they may only interact with each other indirectly through their responses to price signals. Complexity theory, on the other hand, allows for such behaviour as consumers imitating each other's buying habits and producers competing with each other on product quality and other non-price factors.

Over the last fifty years, economics has developed and refined a substantial body of theory which has made it the envy of all the other social sciences. Borrowing from classical nineteenth-century mechanics and from utilitarian psychology, this theory is known to economists as neoclassical equilibrium theory, and it is commonly expressed in the form of linear mathematics. The adoption of a mathematical approach has screened from the public gaze some highly implausible assumptions. It has also meant that every economic system is implicitly treated as if it were a machine, with all the constituent firms and households behaving as if they were pre-programmed robots. Given the necessary information, the behaviour of the whole system and its parts can in principle be predicted precisely, and thereby controlled.

Equilibrium theory might well be appropriate for a planned economy, or perhaps for a very primitive subsistence economy, where little changes, and one year is much like the next. But it is quite unhelpful in understanding how a modern market economy works. Some of the most important elements of the contemporary economy, such as institutions, entrepreneurship and profits, are missing, while some extreme assumptions which contradict experience, such as perfect knowledge, optimising behaviour and constant returns to scale, have been introduced into the theory in order to achieve predictability. The justification of equilibrium theory therefore rests not on its explanatory power but on its predictive ability. On this point the evidence is overwhelming. All surveys of forecasting accuracy come to the same conclusion: economic forecasts based on equilibrium theory perform no better than naive forecasts which assume that next year's values will be the same as this year's.

For most of the period since the Second World War, economic policy throughout the Western world was founded on the implicit belief that national economies were like machines whose performance could be controlled by appropriate government intervention. Governments had at their disposal a battery of regulations, instruments of taxation and monetary controls, as well as direct control of a large part of the economy through public ownership. By pulling on the right levers at the right time with just the right amount of effort, it was thought that whatever results were desired could always be achieved. Thus governments were held to be responsible for the successful performance of their economies. This was a responsibility which politicians were happy to accept, since it appeared to give them greater power. The result was some spectacular and embarrassing failures of policy. Politicians learned from experience that it is unwise to place too much weight on economists' advice. Even predictions of the state of the public finances just twelve months ahead could not be relied upon. As a result, within the last two decades the principal instruments of government control of market economies have been loosened, dismantled and in some countries abandoned altogether.

While economists look at economic activity from the perspective of the economy as a whole, businessmen see exactly the same phenomena from the point of view of their own firm. Unlike politicians, businessmen never accepted the mechanistic theory of how the economy works, and conventional economic theory was not taught in business schools. Indeed, for a long time in most business schools no theory was taught at all. By adopting the case study method, the study of business was deliberately kept free of theory in any form. More recently, a desire for academic respectability has led teachers of business to indulge in theorising. Some business theorists have adopted biological metaphors for business activity. The firm is represented not as a machine but as a living organism. And some economists, such as Marshall, Veblen, Alchian and Boulding have from time to time resorted to biological analogies, applying them to the economy as a whole.

While they are an improvement on mechanical metaphors – an economy is after all a network of relationships among living creatures – biological metaphors also have their limitations. Although human beings are subject to the same influences as other living organisms, unlike other creatures they are capable of making and executing plans, and thus of influencing their environment to an extent which non-humans cannot match. These peculiar human capabilities are demonstrated to the full in the world of economic activity. In order to understand how an economic system works, neither a mechanical nor a biological metaphor is appropriate. What is needed is not a metaphor but a theory, one which is about peculiarly human interactions.

The search for an explanation of how a market economy actually works is one of the oldest quests in economics. In *The Wealth Of Nations* (1776), Adam Smith was the first to offer a convincing solution. He observed that the natural desire of individual human beings to improve their material condition led them to trade with each other. Some people trade the produce of their land, others their capital, but most people trade their labour. Trade leads to the gradual development of specialisation in tasks and functions, and thus is set in motion a process of growth in the economic system as a whole. It is that process of growth which in just over two hundred years has led those countries in which a market economy first developed to evolve from a state in which the great majority of the population lived in abject poverty to an undreamed-of state of affluence at the present time.

The market economy is therefore not a static structure, like a machine, but a process characterised by continuous change. The process normally takes the form of cumulative growth (a virtuous circle), whereby an increase in the size of the market for a particular commodity permits increased specialisation of function, which in turn permits the introduction of new technologies dedicated to the new, more specialised, tasks. The application of the more specialised technologies leads to an increase in productivity, which means higher real incomes for at least some of the participants, whose increased purchasing power expands the size of the market, and so on. The result is increasing returns to scale at the level of the economy as a whole, illustrating the contribution of society to the aggregate dividend.

Of course, this cumulative growth is not a smooth process. On the contrary, it is a highly turbulent one which has aptly been described as 'creative destruction'. The introduction of a new technology renders unprofitable some existing process of production. The introduction of a new product takes away some of the demand for some existing products. In both cases, jobs are lost and the value of the existing plant and equipment is reduced. Although the process of growth creates more jobs in total than it destroys, the new jobs created are unlikely to be in the same place nor are they likely to require the same skills as those which have been lost. And the more advanced the economy, the less likely it is that profitable alternative uses can be found for highly specialised machinery rendered redundant. But although the market process can be greatly disruptive at the level of the individual firm or household, the outcome at the level of the economy as a whole is an orderly and beneficent one. Orderly, in the sense that the process promotes through profit-seeking the continuous adoption of increasingly productive techniques, and beneficial in the sense that the real income of the economy in aggregate is increased.

In addition to the introduction of new products and new

processes, other changes mark the progress of the market economy. New methods of distribution are introduced, new markets are opened up, and new sources of raw materials are discovered. Such changes, with their creative origins and destructive consequences, don't just happen automatically, as they would in a mechanical system. They are brought about by the actions of individual human beings. And not just any human being. To be a successful agent of change, that is, to perceive opportunities for profit which others have not recognised and then to mobilise the resources necessary to exploit them, requires peculiar aptitudes.

So important are the aptitudes required for an agent of change in the process of change which is a market economy, that economists have given them a special name – entrepreneurship – and the agents themselves are known as entrepreneurs. Like many other human aptitudes, these may not be evenly distributed in the population. Most managers in business carry out relatively routine functions within a fairly familiar environment. Entrepreneurs are those businesspeople who are prepared to try new ways of doing things which they hope will make them a profit but which may in fact result in loss. Entrepreneurs proceed not by planning but by trial and error: for them, progress is a question of learning by doing.

The continuing attempts by entrepreneurs to make a profit by adopting new ways of doing things result in a market economy characterised by perpetual novelty. Growth in the market economy is not so much a question of the appearance of more of the same types of commodities, as of new and different commodities being produced, traded and consumed. Perpetual novelty is not just a peripheral feature of the advanced market economy: it is one of its defining characteristics. To mobilise resources, entrepreneurs must interact with other human beings; with their employees, with bankers, with governments, and, above all, with each other in a process of competition. Contrary to the conclusions of equilibrium theory, the outcomes of these interactions are normally unpredictable. This is so for three reasons:

- First, when two human agents interact, the behaviour of one usually depends in large part on what the other will do. In such cases, the relationship is typically non-linear: such an interaction cannot therefore be represented by traditional (linear) mathematics. It can also be shown mathematically that whenever there are small differences in initial conditions between two otherwise similar relationships, then, if the pattern of interaction between the variables concerned is a non-linear one, the outcomes after a small number of time periods have elapsed will be quite different.

- Second, there is the additional factor of chance which enters decisively and frequently into human relations. A businessman who misses his flight may thereby lose an important contract, changing the whole future of his business and the subsequent lives of his employees.
- Third, there is free will: individual human beings are free to choose, and we cannot always tell in advance how they will react. Contrary to the belief of advocates of rational decision theory, at critical moments important decisions may not always be rational. For example, most private investors buy near the top of the stock market and sell near the bottom, which is a very irrational thing to do.

To summarise, the behaviour of a market economy exhibits certain distinctive characteristics: first of all, it is a process not a structure. It is the result of the activities of millions of individual human beings interacting with each other through the processes of production, trade and consumption. While each individual pursues independently his or her own objectives, and while outcomes are frequently disruptive at the level of the individual household or firm, the overall outcome is both orderly and beneficial at the level of the economy as a whole. This order or pattern which emerges was neither intended nor even anticipated by the participating individuals. Like the emergence in history of such other institutions of human society as the law, language and money, it may be said to be the result of human action, but not of human design. Its emergence is an example of the phenomenon of self-organisation.

The market economy is also characterised by adaptation to a changing environment. The agents of adaptation are the entrepreneurs and their environment includes the behaviour of competing entrepreneurs. For entrepreneurs, competition is a process in which they acquire new knowledge. The exploitation of this new knowledge results in perpetual novelty in the economy as a whole. Although the outcomes of market processes are unpredictable in detail, they do exhibit a regularity of pattern which is to a certain extent predictable.

When taken together, these features of self-organisation, adaptation, perpetual novelty and regularity of pattern but not of detail, exhibited by market economies are recognisable as characteristic features of a class of processes known as complex adaptive systems. Complex adaptive systems are to be found amongst biological, chemical and artificial (computer-generated) phenomena, as well as amongst human phenomena, and their properties are the subject of intensive investigation by both natural and social scientists at the present time. For those who are rightly wary of drawing analogies between human behaviour and other classes of phenomena, it must be emphasised that in discussing complex adaptive systems we are not dealing with analogies but with the *same* processes at different levels of complexity. It is helpful to think in terms of a hierarchy of complexity amongst phenomena. At the lowest level, physical phenomena obey the relatively simple laws of matter. At higher levels, chemical, biological and then human phenomena follow increasingly complex patterns of behaviour.

Recognising the market economy as an example of a human complex adaptive system improves our understanding of how a market economy works, compared to equilibrium analysis. Job insecurity, for example, is an issue of increasing importance in the most advanced market economies. Frequent changes of employment mean that an increasing number of workers may evolve into a kind of self-employed status. For those who are risk-averse, job insecurity raises the question of whether governments should become employers of last resort. Neoclassical theory, which assumes the economy to move effortlessly from one equilibrium to another, is simply not equipped to deal with questions of adjustment. On the other hand, the complexity approach focuses on the process of adjustment. It suggests that, in general, the greater the degree of complexity, the greater the costs of adjustment are likely to be. Likewise, equilibrium theory is of little assistance to politicians and civil servants trying to decide what should be the objectives of market regulation and what constitutes anti-competitive behaviour.

The application of complexity theory suggests that growth and fluctuations in market economies may be inseparable. Claims of politicians to be able to eliminate the business cycle should therefore be treated with caution. So long as governments lack the knowledge to be able to fine-tune policy to cope with fluctuations in the economy, the implication for policy is that governments should manage their own finances so as to be able to borrow heavily when a recession occurs.

Among the other implications of the complexity approach is that while pattern predictions are often possible, economic forecasting of the traditional kind implies a degree of precision that is quite unattainable, no matter how well-informed is the forecaster. Which is why official forecasts of the public finances are invariably wrong, and why, for example, the members of the Monetary Policy Committee of the Bank of England usually disagree about the desirable change in interest rates. While we can predict with some degree of confidence that an expansion in the supply of money will normally lead to an increase in the rate of inflation, we cannot say in a particular case when that will happen nor by how much it will increase.

A further advantage of the complexity paradigm is that it provides a common mode of discourse between theories of business and theories of economics. When applied to business, complexity theory explains why 'muddling through' is a process more commonly encountered in managing a company than the deliberative process founded on rational calculation imagined by many equilibrium theorists. However, complexity does not deny the possibility of a company choosing and following a strategic direction. It therefore takes a different view of strategy than that of business theorists such as Mintzberg and De Geus, and others who believe that businesses should respond to events in a purely reactive way and that senior managers should refrain from trying to set and pursue strategic objectives. Indeed, complexity theory lends support to one particular model for the organisation of the modern firm. In this view, senior managers should confine themselves to handing down broad objectives and guiding principles, while decisions on implementation and adjustment should be devolved to self-organising business units. The alternative mechanical model of control whereby there is rigid adherence at lower levels to detailed rules of procedure laid down from the top is generally likely to be less effective, as the performance of most publicly-owned organisations in a market economy illustrates.

In terms of economic theory, the proposed approach is congruent with both classical and Austrian perspectives. In contrast to neoclassical equilibrium theory, they see the market economy as a process in which adaptation to new knowledge takes place, and where institutions evolve. Marx and Schumpeter, for different reasons, believed that the institutions of capitalism would inevitably evolve into socialism. Today, Gray and others suggest that the market process is undermining the cohesion of societies in a way which will lead eventually to anarchy and war. But these outcomes, too, are far from being inevitable. In human complex systems like the contemporary advanced market economy, those institutions which have emerged as the result of unintended human actions exist side by side with those which are the result of deliberate human planning. In the future, it seems likely that the latter type of institution will play an increasingly important part in the market economy as well as in human affairs generally.

In Chapter 2, the dominant paradigm in contemporary economic thought, neoclassical equilibrium theory, is shown to treat the market economy as if it were a machine. As a result, outcomes are predictable, and achievable through government control. The development of neoclassical theory in the postwar period is outlined in Chapter 3. It is argued that excessive refinement has carried the theory beyond the point where it can serve as a useful method of analysis. Chapter 4 discusses the way in which the mechanical metaphor laid the foundation for a greatly expanded role for government intervention in the market economy. But experience has revealed its limitations, so that the making of policy has now become uncoupled from equilibrium theory.

Chapter 5 explores the application of equilibrium theory to business, and explains why the study of economics and the study of business have followed such methodologically divergent paths. Business theorists have sometimes been tempted by biological analogies, and Chapter 6 explores the relationship between biological and mechanical analogies. While the former generally represent an improvement on the latter, they still fall short of a satisfactory theory, since they do not account for the distinctly human element in economic activity. The concept of complexity is introduced in Chapter 7. The chapter begins with a theory of complex phenomena, describes some properties of complex adaptive systems, and outlines the characteristic features of an economy represented as a human complex adaptive system. The remainder of the book explains how different aspects of a market economy work, as seen from the perspective of complexity. Chapter 8 shows how the market economy resolves the problem of coordinating the wants and expectations of participants. Chapter 9 discusses the evolution of the institutions of a market economy, while Chapter 10 examines what lessons can be learned from the study of economic history. Complex systems result in outcomes in the form of persistent and recurring patterns, and three such patterns are discussed in Chapter 11: the business cycle, patterns in the location of economic activity and regularities in the relationships between industries. Chapter 12 looks at the functions of entrepreneurship and corporate organisation, while Chapter 13 discusses the implications of complexity for economic theory. Chapter 14 examines the implications for business strategy and government policy-making. Chapter 15 considers the likely future evolution of the market economy, and the Conclusions of the book are presented in Chapter 16.

2
The Economy as a Machine

'Systems which have universally owed their origin to the
lucubrations of those who were acquainted with one art,
but ignorant of the other; who therefore explained to
themselves the phenomena, in that which was strange
to them, by those in that which was familiar; and with
whom, upon that account, the analogy, which in other
writers gives occasion to a few ingenious similitudes,
became the great hinge on which everything turned.'

Adam Smith, *Essay on the History of Astronomy*

'You treat world history as a mathematician does mathematics, in which nothing but laws and formulas exist,
no reality, no good and evil, no time, no yesterday, no
tomorrow, nothing but an eternal shallow mathematical
present.'

Hermann Hesse, *Das Glasperlenspiel*

When Adam Smith, the founding father of modern economics,
published his great treatise *The Wealth of Nations* in 1776, he
understood that the market economy represented a network of human
interactions. Today, modern economic theory has been purged of
all human content, with the result that it is able to offer only a
very limited and quite distorted picture of how a contemporary
market economy works. Where did economic theory go astray?

13

It happened when the theory of value, that is the theory that attempts to explain what determines relative prices and quantities in a market economy, took over as the main line of enquiry from the theory of growth. When that happened in the first half of the twentieth century, attention was switched from the creative to the allocative functions of the market, to use Kaldor's phrase.[1] Although both allocative and creative elements are to be found in *The Wealth of Nations*, the primacy of the latter in Smith's thinking was clearly indicated by the full title of his book: *An Inquiry into the Nature and Causes of the Wealth of Nations.*

Despite advances made in the theory of value towards the end of the nineteenth century, the theory of economic growth remained of primary importance up to the time of Alfred Marshall (1842–1924). Marshall, who had published the first edition of his *Principles of Economics* in 1890, was responsible for establishing economics at Cambridge as a legitimate and independent field of scholarship. The subsequent emergence of economics as a professional discipline can be attributed in large part to his influence. For Marshall, as for Adam Smith, the main purpose of economic theory was to investigate the fundamental causes of economic growth – why do some societies grow rich while others remain poor? The method of analysis he proposed to try to answer this question incorporated important elements drawn from physics. In particular, the notion of equilibrium, which was to become central to conventional economic thought in the twentieth century, was borrowed from classical mechanics. Indeed, the method of economic theorising introduced by Marshall and now known as neoclassical economics has been well-described as a cross between utilitarian psychology and differential calculus.

But while borrowing methods of analysis from physics gave the subject that apparent scientific respectability which academic acceptance of a new professional discipline required, it also created a dilemma. Marshall recognised that the principal characteristic of economic activity was progress or change. In the Introduction to his *Principles* he wrote that the central idea of economics must be that of living force and movement, and therefore that the methods of analysis most appropriate for economics should be drawn not from physics but from biology.[2] But, since biological conceptions were 'too complex' for laying the foundations of the subject, Marshall suggested it was permissible for economists to begin with what he called 'mechanical analogies'. But he warned that 'the limitations of mechanical analogies and of the statical theory of equilibrium are so constantly overlooked especially by those who approach it from an abstract point of view, that there is a danger in throwing it into a definite form at all'.[3]

Unfortunately, conventional economic thought from Marshall's death down to the present day completely ignored his warnings. Instead, it continued to refine and develop the mechanical analogy. By excluding all those factors in economic life that did not lend themselves easily to measurement, and by making drastic simplifications of real world relationships, an entire modern economy came to be represented in economic theory by nothing more than a set of linear equations.

Amongst the factors excluded from the picture of economic activity represented by these equations were all organisations such as governments, all institutions such as the law or social class, and all the 'non-rational' elements of human behaviour like confidence, trust, ambition, benevolence and malice. Even such factors as profit, innovation and enterprise, which most observers of economic activity would think of as being amongst the principal characteristic features of a modern market economy, were omitted. Amongst the simplifying assumptions, the most important one made was about knowledge. Every firm and every household was assumed to have perfect knowledge not only of all existing prices and opportunities in the economy, but of all future prices and opportunities as well.

The neoclassical economists who developed equilibrium theory after Marshall seemed to overlook the fact that economic activity is an historical process involving interactions between human beings. Instead, they proposed that it should be seen as a mechanical procedure, automatic in response, repetitive in form, reversible in principle and precisely predictable in outcome. This view of economic activity gradually grew in influence until it dominated the economics profession. It is known as general equilibrium theory. It is an equilibrium theory in the sense that supply and demand must at all times be in balance for every good and service traded in the economy, and it is general in the sense that all markets are interrelated and must balance simultaneously. A change in price in one market feeds through into prices in other markets. So complete has been the adoption of general equilibrium theory amongst academic economists throughout the world that it is now regarded as the only acceptable starting point for any logically consistent explanation of the behaviour of market economies. In the rest of this book we shall refer to it as equilibrium theory or the mechanical model of the economy.

General equilibrium theory represents the economy as a mechanical system in which at any moment in time there are fixed and variable elements. The fixed elements include consumer preferences, technology and the available supply of resources. The variable elements are the prices and the quantities of each of the commodities produced and consumed in the system. Given any initial conditions, the system will instantly converge to a unique point representing that set of prices which will clear all markets simultaneously. This set of prices and its associated quantities produced and consumed will remain unchanged until there is some further change in initial conditions.

The theory does not pretend to offer an explanation of how market prices are actually determined in the real world (although some economists seem to believe that it does). Rather, the theory is a set of propositions logically deduced from a number of basic assumptions or axioms, invariably formulated in terms of linear algebra. The development of the theory over the past half century has been directed to finding the minimum number of assumptions necessary to prove the existence, uniqueness and stability of a set of equilibrium prices, that is those prices which would clear all markets simultaneously. No attempt has been made to see whether this set of prices bears any relation to actual prices, nor whether the theory has any explanatory power, either in the sense of a capacity to predict or in the sense of offering an improved understanding of how a market economy actually works. Nor has there been any attempt to verify the realism of the basic assumptions.

The method of reasoning employed by the neoclassical school is essentially the same as the procedure followed by Ricardo as described by Schumpeter:

This is the method of analysis which proceeds by excluding as many variables as possible. Then, for the rest, piling one simplifying assumption upon another until, having really settled everything by these assumptions, he [Ricardo] was left with only a few aggregative variables between which, given these assumptions, he set up simple one-way relations so that in the end, the desired results emerged almost as tautologies.[4]

Nevertheless, it was hailed by almost all professional economists as being a triumph for economic theory when in 1964 it was shown that, given specified endowments of resources, technology and consumer preferences, a system of competitive markets could in a logical sense be proved to exist. This triumph was only slightly marred by the fact that its proponents themselves were divided on its interpretation. Since it was recognised that the result had been achieved by making such drastic simplifications and such sweeping exclusions as were not remotely attainable in practice, did it mean that a real world market economy could work, or did it mean that it could not work?

It is not difficult to see the machine-like characteristics of the system. Given values for the fixed elements and for the initial conditions, outcomes are uniquely and precisely predictable. In principle, the results are repeatable, and the processes are reversible: they are independent of history. There is the attractive possibility of control and direction by a human being who may be supposed to stand outside the system, perhaps an economist hoping to bring about a better world or a politician hoping to get re-elected. Provided the right levers are pulled (that is the correct initial conditions are set), the system will always deliver the desired results.

The mechanical properties of this theory are more than a figure of speech. In the 1950s an Australian economist called A.W. Phillips, who had trained as an engineer, constructed a hydraulic machine to represent the flows of expenditure through the different sectors of a market economy, based on the principles of equilibrium. Several of these machines were built, and one now rests in a basement in the London School of Economics. While the idea of the economy as a machine may be acceptable for teaching an introductory course in economics, it is quite inadequate for a serious understanding of how a contemporary market economy works, and it certainly provides an inappropriate basis for formulating economic policy. In addition to perfect knowledge, other assumptions are made in order to fit the mathematics which are contrary to the experience of the world. These include the assumptions that it would be impossible for a firm to cut its unit costs by increasing its output (that is no increasing return to scale), that the buying and selling decisions of any one firm cannot significantly affect the market price of a commodity (perfect competition), that the range of available goods and services and the processes for their production and distribution remain unchanged, and that all markets clear smoothly and instantaneously. The existence of any unemployment anywhere at any time is a straightforward contradiction of the last assumption. It is also assumed that market relationships are wholly impersonal, and that flows of information within the market economy exclusively concern prices.

The machine metaphor not only leaves out some aspects of human nature which are essential ingredients of business performance, like confidence, imagination and ambition – but even more important phenomena like profit, enterprise, and innovation (that is the introduction of new goods and services and new ways of making and distributing them). Above all, it leaves out chance or uncertainty. Since all of these things constitute the central features of a market economy, it is not surprising that neoclassical or equilibrium economics has failed to convince anyone outside the academic branch of the economics profession of its efficacy.

The fundamental flaw in the metaphor of the economy as a machine is the assumption of perfect knowledge. The successful performance of a machine depends on the availability to its controller or engineer of complete knowledge of every aspect of its operation. But in every developed market economy knowledge is scattered and specialised. Different people know different things: a chef knows more about how to prepare a range of chicken dishes than an electrician does. Even if such dispersed information could somehow be collected, there would still remain the problem of tacit knowledge. All of us have acquired through experience much more knowledge than we can ever articulate.

And even if the problem of tacit knowledge could somehow be overcome, there would remain the still more fundamental problem of our lack of knowledge about the future. People acquire knowledge as they act. So long as changes in an economy are taking place spontaneously at the level of the individual, then the future of that economy cannot be known, and without knowledge of the future, a market economy cannot be supposed to operate like a machine.

The machine might seem to be an appropriate model for a centrally planned economy if not for a market economy. But one of the principal reasons for the failure of the experiments in central planning which took place in the Soviet Union and elsewhere was that it proved to be impossible for all the relevant information necessary for effective decision-making to be collected and communicated to one central point. And even the almost unlimited capacity of today's computers to process information would not be able to overcome the problem of its acquisition.

It is possible that the machine metaphor might be appropriate in a primitive subsistence economy cut off from the outside world. Here, the processes of production and consumption of an unchanging set of commodities using traditional practices might repeat themselves from one year to another. Where the future can be counted on to be much the same as the past, it might be acceptable to treat the economic system as if it were a machine. But it is a quite inappropriate metaphor for a modern market economy.

The existence of uncertainty in economic affairs is not denied by exponents of equilibrium theory: it is simply dismissed as being of little practical importance. The analogy is usually drawn with the calculation in physics of the velocity of a falling body. This velocity can only be calculated with accuracy if the body is assumed to be falling through a vacuum. In practice, for most bodies falling through the air, the actual velocity will not be very different from that which would have been observed in a vacuum. In the same way, it is argued, the behaviour of individual agents in an actual market economy is not very different from that which would have occurred if there were perfect knowledge rather than uncertainty. But if the world were really the world of perfect knowledge and foresight assumed by equilibrium theory, the buying and selling decisions of every businessman and of every housewife could be calculated by a computer. There would be no need for risk-taking, nor for profit-seeking behaviour.

Once it is recognised that, whether we act in our capacity as consumers or producers, our knowledge of the future is always going to be fragmentary, incomplete and quite often mistaken, then the role of the enterprising businessman willing to take risks in pursuit of a profit (which he may or may not realise) comes into its own. In an actual market economy, competition is not just a repetitive mechanical process but it becomes a discovery procedure in which new knowledge is discovered and communicated.[5]

Just as a machine cannot be conceived of as experiencing spontaneous change within its component parts, so equilibrium theory cannot cope with changes originating spontaneously within the economic system. This explains why some economists are always talking about *external* shocks to the system. In practice, external shocks are infrequent and relatively unimportant but neoclassical economists focus on them because their equilibrium theory can, at least in principle, try to analyse their effects. But the theory is helpless to deal with the much more important and more frequent changes which arise from within the market economy – the *internal* shocks. One of the characteristic features of modern market economies is incessant change. New consumer and intermediate goods, new types of plant and equipment, new services and new methods of producing and distributing them are continuously being created, and old ones are being destroyed as part of the process of market competition. But because there is no place in equilibrium theory for internally created changes, there is therefore no role for entrepreneurs nor for profits nor innovation, and there is no scope to consider questions of adjustment.

And what about human behaviour? Having made the basic assumption of perfect knowledge, the second step which equilibrium theory takes towards achieving predictability is to make some special assumptions about individual human behaviour. It is assumed that human beings will in all circumstances act consistently, aided by perfect knowledge, in the pursuit of certain clearly defined objectives. As consumers, people will always arrange their purchases so as to maximise their individual satisfaction. As workers, people will seek always to balance at the margin their income with the dissatisfaction of the job. As employers, they will always try to maximise their profits. This is known as 'rational' behaviour.

One of the most misleading assumptions of equilibrium theory, reflecting its mechanical origins, is that human beings in their interactions with one another behave no differently than do such inanimate objects as billiard balls. Not only are human beings assumed to be entirely passive in their behaviour, returning to an inert state after some disturbance, but they are assumed not to adjust their behaviour as a result of their experiences of interacting with their environment. No attempt has been made to discover whether people in fact do behave like this. Evidence of apparently irrational behaviour, such as smoking, drinking, drug-taking, driving at high speed or speculating on the stock exchange is brushed aside. Keynes wrote of those who invested in the stock market as being driven by 'animal spirits', and there is certainly evidence of many investors buying near the top of the market, and selling near the bottom. And the abundant evidence of herd-like behaviour in purchasing decisions suggests that consumers do in fact adjust their behaviour as a result of their interactions with each other.

The problem of verification of the theoretical structure of equilibrium has been largely ignored by the academic profession, the great majority of whom have accepted its conclusions as if they were a description of reality. Whereas in a science like physics, observations which cannot be made consistent with existing hypotheses lead to a reconsideration of the basic assumptions or axioms of the theory, in economics, observations which contradict the basic hypotheses of general equilibrium theory have simply been disregarded. In most scientific theories, the basic assumptions are selected by observing the phenomena whose behaviour forms the subject matter of the theory. But the basic assumptions of general equilibrium theory are either unverifiable, for example that consumers maximise their satisfaction, or are directly contradicted by observation and experience.[6]

During the postwar development of neoclassical theory all such considerations were cast aside in the pursuit of predictability, believed to be the hallmark of a science. Those who constructed models of the economy based on equilibrium theory were elated at the prospect of being able to provide answers to questions such as the following: If the wages of coal miners rise by £10 per week, what will be the effect on the price of a loaf of bread? If short term interest rates rise by half a percentage point in the next twelve months, by how much will employment fall?

The apparent capability of producing exact answers to questions like these seemed to put economists on a par with natural scientists, and therefore a cut above their less fortunate brethren in the 'softer' social sciences like history or sociology, whose answers were always ambiguous. Of equal importance was the fact that equilibrium theory seemed to offer the prospect that economists in government service would be able to control the performance of a market economy in much the same way as engineers could control a machine. By calculating the size of the necessary policy measures, they could bring about whatever outcome was wanted. Full employment? No problem. As soon as we have constructed and tested our model, we will let you know what policy changes to make.

It is difficult at the present time to imagine that frisson of excitement which economists felt fifty years ago at the prospect of achieving the status and esteem then accorded to natural scientists. Not only could they come up with precise answers to practical questions, but the equilibrium theory they had developed to allow them to do this was expressed in mathematical form. Although they could not understand what these equations meant, politicians, businessmen, and other lay people were nevertheless able to recognise that the use of mathematics was characteristic of true scientific endeavour.

The use of mathematics concealed from those same politicians and businessmen the fact that these exact answers were produced as a result of making some very implausible assumptions. Had these assumptions been spelled out in plain English, the answers might have been received with a greater degree of scepticism. How could such extreme assumptions as *perfect knowledge, ra-*

tional behaviour, perfect competition, constant returns to scale and so on possibly be justified? Those who were responsible for the development of equilibrium theory were of course aware of the unrealistic nature of their assumptions, and two justifications were offered. To begin with, the assumptions were justified as being purely provisional. It was hoped and expected that further refinement of the theory in the future would lead to their gradual relaxation and eventual abandonment. The analogy which was suggested was with the erection of scaffolding as a preliminary to the construction of a building. In fact, equilibrium theorists long ago lost any interest they may originally have had in the building, and have become entirely pre-occupied with the development of their scaffolding. As we shall see in the next chapter, with every reformulation of the theory, the scaffolding has become thicker and more impenetrable, with growing doubts as to whether there is a solid building underneath.[7]

The second justification was the philosophical argument that in constructing a theory the realism of its assumptions was unimportant. The only thing which mattered was the ability of the theory to predict successfully. We shall see in the next chapter just how unsuccessful predictions grounded in equilibrium theory have been in practice. Meanwhile, it should be noted that the criteria for judging a theory should include its explanatory power, that is to say its consistency with informed judgements as to what is really going on. In Nelson's phrase, the explanation offered by a theory should 'ring right' to those who know the details of the field.[8] As a consequence of the unrealistic nature of its assumptions, neoclassical equilibrium theory has no explanatory power. It contributes nothing to our understanding of how a modern market economy works. Indeed, so restrictive are its assumptions that it might even be said to subtract from our understanding of how a market economy actually works.

Not all economists were carried away by the tide of equilibrium theory which swept through the profession in the second half of the twentieth century. A tiny handful of dissenters, known as Austrians because they followed the teachings of the late nineteenth-century Austrian economist Carl Menger, refused to accept equilibrium theory. One of their leading members, Friedrich von Hayek, wrote a series of articles during the Second World

War which questioned the foundations of equilibrium theory.[9] He pointed out that to apply unthinkingly to a social science like economics, which dealt with the behaviour of human beings, methods of analysis drawn from physics, which dealt with inanimate objects, was itself an unscientific procedure. He warned that it was particularly foolish to suppose that future economic events could be predicted as if the economic system behaved like a machine. At this time, Hayek was already a respected figure in the profession, with an established reputation as a scholar and a Chair at the London School of Economics. Yet his warnings, like those of Marshall thirty years earlier, were completely ignored.

Instead, academic economists after the Second World War flocked to embrace the new 'scientific' economics. Not only did it give them apparent parity of esteem with their colleagues in the natural sciences, it satisfied their emotional needs. Like doctors or engineers, they could actually do something to better the lot of everyone in society. They could enter government, and improve the management of the nation's affairs, in particular by eliminating unemployment. As a result, lavish research funds flowed in from grateful politicians; Nobel prizes were established and distributed; and the stage was set for the spectacular failure of equilibrium economics in the postwar world. To achieve outcomes which are precisely predictable, it is necessary to assume that the domain of economic activity is both certain and linear. In fact, as all practitioners know only too well, it is uncertain and non-linear.

3
The Pretence of Knowledge

'To accommodate assumptions not to reality but to the requirements of technique will be thought a rather dubious scientific procedure'.

J.K. Galbraith, *The New Industrial State*

As we saw in the last chapter, equilibrium economists do not pretend that their theoretical assumptions are descriptively accurate. They admit that their theories do not offer an explanation of how people in a market economy actually behave, but they say this does not matter. Their argument runs as follows: The only valid test of a scientific theory, they assert, is the accuracy of its predictions. The realism of its assumptions are quite irrelevant. Take, for example, the case of someone playing a game of snooker. A good player will always strike the cue ball as if he understood the physical laws of motion which govern the direction and speed of the ball he is attempting to sink in a pocket. If the ball actually goes into the pocket he intended, then he has acted as if he had correctly made the relevant calculations, even though we know he is entirely unaware of what these might be. In exactly the same way, it is claimed, people engaged in economic activity in a market economy, whether as producers or consumers, act as if they were following the laws of behaviour laid down by general equilibrium theory, even though they are blissfully unaware of doing so.

Let us accept for the moment the proposition that the only

valid test of a theory is its ability to make accurate predictions. No-one would dispute that the laws of motion which govern the behaviour of snooker balls have predictive accuracy. They can be verified by experiment, and are demonstrated by daily experience. But what about the validity of the equations of general equilibrium, which allegedly underlie human behaviour in the market economy? These equations are much too general to be tested directly, and of course experiments with real people are ruled out. But general equilibrium theory has given rise to many hundreds of subsidiary models whose predictive ability is tested regularly. These are the models which have been devised by economists with the specific intention of forecasting some aspects of the performance of particular market economies.

Some of these models are very large, up to a thousand equations is not unknown; others are quite small, perhaps composed of no more than three equations. Some forecast up to five years ahead, but most venture no further than the next four or six quarters. Some produce forecasts quarterly, others annually. What they all have in common is that when subjected to the simplest tests of forecasting accuracy they fail. Although it is unusual in economics to be able to make general statements without qualification, it has been shown quite conclusively and unambiguously that predictions based upon equilibrium theories have invariably turned out to be wrong, usually by large margins.

The failure of economic forecasting

In the United States and the United Kingdom and throughout the industrialised world, there are literally hundreds of organisations providing regular economic forecasts. Many have been doing so continuously for at least 25 years, so that there has been ample opportunity to test their forecasting accuracy. Some of these organisations are governmental, such as the Federal Reserve Board and the President's Council of Economic Advisers in the USA, or the Treasury and the Bank of England in the UK. Others are international organisations like the IMF or the OECD, but most are private commercial organisations selling their forecasts to clients. Typically, they attempt to forecast a limited range of selected economic variables, including annual changes in aggregate national output (GDP or GNP) and in the aggregate price level (the rate of inflation).

Over the past three decades there have been periodic assessments of the accuracy of such forecasts, although no-one seems to have paid much attention to their results. From time to time these assessments have been reviewed, and the findings of the reviewers published in books and articles, which again have attracted surprisingly little attention. The findings are invariably the same, and may be summarised in the following propositions drawn from Sherden's comprehensive survey[1] of the record of economic forecasting:

1. *Economic forecasts have generally failed to predict turning points in the behaviour of the economy* (that is, the timing of the peaks and troughs of the business cycle). The timing of turning points in the business cycle is the kind of information which both policy-makers and market participants would like to have above all else. Zarnowitz analysed the error rates of the six major US forecasting organisations in predicting GNP and inflation between 1970 and 1980, a period of some volatility. He found that of the 48 forecasts he studied, 46 missed the turning points. Likewise, another tracker of economic forecasts, McNees, who studied the forecasting record of the Federal Reserve Board from 1980 to 1995 found that the Fed correctly predicted only 38 per cent of the turning points in GNP growth and inflation during that period. This is a worse record than could have been achieved by tossing a coin.

2. *In general, economic forecasts perform no better than guessing.* As an illustration of just how bad economic forecasts are, one can look at the forecasts made by the governments of the seven major world economies over the period 1987–92. In a comparative survey undertaken by the OECD, the official annual forecasts of inflation and of the growth of output in each country 12 months ahead were compared with actual outcomes.[2] The benchmark used for comparison was the simplistic projection that next year's output growth or inflation rate would be the same as this year's. For the forecasts of output, this simple rule performed at least as well as the official forecasts. In predicting inflation, it performed slightly better than the official forecasts. These results are quite typical of

the performance of national government forecasts in general.
Perhaps private forecasters, spurred on by the prospect of
gain, could do better? On 28 December 1996, the *Financial
Times* published the results of forecasts made by six leading
City of London financial forecasters 12 months earlier. They
had then been asked to forecast the year-end levels of four
stockmarket indices, four rates of interest, the consumer price
index, the pound/dollar exchange rate, and the prices of gold
and oil. The average error rate of the six City forecasters ranged
from 9.2 per cent to 17.8 per cent. But a naive prediction
based on the rule that the future would be much like the past
(for example if it had been assumed that end-1996 interest
rates would be the same as the end-1995 ones), would have
produced an average error rate of only 8.5 per cent. When
the same comparison was made over the year 1995, the average errors of the panel of six experts ranged from 11.8 per
cent to 20.3 per cent. The naive forecast had an average error
rate of just over 10 per cent.

Reviewing all the evidence of tests of forecasting accuracy, Sherden
concludes that while the forecasters are able to predict movements in rather stable variables like government spending with
greater accuracy than the naive forecast, they are less successful
at predicting the more volatile and more interesting variables
like interest rates. For the most frequently forecast variables, real
GNP growth and the rate of inflation, the economic forecasts
turn out to be no better than the naive forecasts.

Among the other conclusions which emerge from Sherden's
survey are that no individual forecaster or forecasting organisation can outperform their peers for very long. This suggests that
any outperformance observed over a short period of time is due
to chance rather than to skill. It also turns out that forecasts
based on more sophisticated methods perform no better than
those based on subjective judgement, while large economic models
perform no better than small ones. Accordingly, it is not surprising to find that the statistical evidence does not support the
belief that there has been any improvement in forecasting success over the past 30 years.

Do economists forecast more accurately than those without
an economics training? In 1985, *The Economist* newspaper con-

ducted a survey amongst representatives of different occupations to see who could predict most accurately the state of the British economy in 10 years' time. The results were published in its issue of 3 June 1995. A group of dustbinmen (garbage collectors) tied for first place with a group of chairmen of multinational companies. It may be of some small comfort for economists to know that when the forecasting record of Wall Street analysts and gurus is examined, whether predicting the movement of the Stock Market as a whole or selecting individual stocks, they too perform badly. So do the managers of mutual funds.[3]

Despite having argued all along that the acid test of a scientific theory was its ability to predict successfully, proponents of the equilibrium theory have apparently remained unmoved by such results. So far from relaxing or abandoning their more unrealistic assumptions, or attempting to reintroduce those important elements which they had omitted, equilibrium theorists have proceeded to still further refinements of their models. As the level of refinement has increased, that is as their models became more abstract and more 'rigorous', the unrealism of the assumptions has grown in order that the theory should fit the more sophisticated although inappropriate mathematics. In 'improving' the equilibrium theory of how the economy worked, logical elegance and mathematical rigour were the only criteria that mattered. Relevance to understanding the real world or even usefulness to government or business did not count. Economists became the prisoners of the models they had built. Whereas normal science aims at understanding the behaviour of a group of related phenomena from the physical world or from the world of nature, economics became preoccupied with an understanding of the behaviour of its own mathematical models – and most of the models bore no relation to any state of the world which has ever existed or is ever likely to exist.

The result of some thirty years' work in this direction was the construction of an imposing edifice of economic theory, expressed entirely in mathematical equations, based on some quite fanciful assumptions. It is rather like a baroque palace made of icing sugar which has been erected on a foundation of sand. It is difficult to create, magnificent to behold, and quite useless for any purpose. Up to about the middle of the twentieth century, when almost all economists still wrote in plain English, readers could judge for themselves whether economic theories made sense. Nowadays, a manifest absurdity can be disguised in mathematical language and applauded as a profound truth. So-called 'real business cycle' theory is a good example. This briefly fashionable theory led believers to the conclusion that unemployment is to be explained by workers voluntarily withholding their labour in anticipation of a future increase in wages.

Rational expectations

That expectations play an important part in human behaviour has long been recognised by economists. But how do individuals actually form their expectations? The way in which equilibrium theory has approached this question is a perfect illustration of its tendency to make arbitrary assumptions rather than choosing assumptions which reflect observed behaviour. In the case of the celebrated theory of 'rational expectations', introduced by Robert Lucas of the University of Chicago in 1976, the assumptions were carefully chosen so that they would be consistent with the already existing equilibrium theory. This is rather like choosing a new piece of scaffolding to fit the existing scaffolding rather than choosing a piece which will support the structure which is to be built.

In the case of rational expectations theory, the longstanding assumption of equilibrium theory that all firms and households are at all times perfectly informed about all present and future prices was augmented by the even more remarkable assumption that all participants in a market economy, from housewives to property developers, have a perfect understanding of equilibrium theory so that they can correctly anticipate the future consequences of present events, and form their expectations accordingly. Of course, this begs the question of whether there is a general belief in the existence of one true equilibrium theory and its application. The fact that different forecasting organisations with access to broadly the same set of information regularly produce forecasts which are quite different from each other suggests that there is not.

Not surprisingly, it has often proved difficult to reconcile rational expectations theory with real world behaviour. If the theory were correct, it would imply steadily rising prices for exhaustible resources, yet observed long-term trends have been remarkably flat. Nor is it possible to account for the actual course of share prices with any form of rational expectations. Neither the excess return observed on equities over bonds, nor the recorded variations in the volume of trading in securities nor volumes of trading in other financial markets can be explained in these terms.

Financial markets are among those where expectations play a particularly important part in the calculations of participants. Yet it is precisely in these markets that rational behaviour is notable for its absence. In Sherden's phrase, every stockmarket is 'a psychological soup'[4] of human emotion, notably greed, fear, hope and superstition, and it is these non-rational factors which probably account for much of the volatility of stock prices. Some equilibrium theorists have gone so far as to deny the possibility of any irrational behaviour in financial markets at any time. For example, Milton Friedman has denied that speculative behaviour can ever be destabilising, an assertion which appears to fly in the face of the historical experience.[5] This suggests that the periodic financial crises which have occurred throughout the Western world in the past four centuries have been marked by herd-like behaviour on the part of market participants, who are gripped by alternating moods of unwarranted optimism and pessimism.

There is indeed a considerable body of evidence suggesting that for many individuals who engage in financial transactions of a speculative nature, emotion is a more powerful influence in their decisions than reason, and that amongst those emotions, hope is stronger than fear, leading to overconfidence. Adam Smith observed that 'the chance of gain is by every man more or less over-valued, and the chance of loss is by most men under-valued'.[6] The continuing popularity of lotteries, where the expected value of a win is always less than the price of a ticket, is a demonstration of this principle. And most if not all professional traders in the stockmarket would agree with Warren Buffet's observation that fear and greed move stock prices above and below a company's intrinsic value. Or as a trader on the New York Stock Exchange more than seventy years ago remarked: 'The principles of successful stock exchange speculation are based on the supposition that people will continue in the future to make the mistakes they have made in the past'.[7]

Rational expectations theory will have none of this. According to the hypothesis of 'efficient' financial markets, all investors share identical expectations ('rationally' formed) of the future prices of assets. All new information that comes into the market is instantly and accurately discounted into prices. Accordingly, sudden and temporary movements in share prices reflect purely rational changes in the valuation of assets rather than shifts in investor sentiment. In this view, the market is wholly rational and mechanistic. It follows from this theory that trading volumes in financial markets are low or zero, and that indices of trading volume and price volatility should not be serially correlated. Unfortunately for rational expectations, data on financial markets transactions are abundant in quantity and of high quality. They demonstrate conclusively that trading volumes and price volatility are large, and both series show significant auto-correlation.[8] And few people believe that the crash of the New York stockmarket in 1987 can be attributed to new market information being issued at the time.

In response to these criticisms, some behavioural or 'noise trading' theorists have built models of stockmarket behaviour in which there are two classes of trader, 'irrational' investors and 'rational expectations' traders. Such models suffer from the defect of all models with 'rational expectations' ingredients, namely how to explain *how* the 'rational' traders acquire their full knowledge of the expectations of others. A further weakness of these models is that they assume that the 'irrational' traders never discover eventually that their expectations are false.

A much more satisfying approach to understanding how financial markets really operate is the complexity theory approach proposed by Arthur, Holland, LeBaron, Palmer and Tayler.[9] It is more satisfying, not only because the assumptions do less violence to our understanding of how individuals do actually form their expectations, but because the outcomes are consistent with some observed statistical patterns of financial-market behaviour. Arthur *et al.* assume that markets are made up of people ('agents') each of whom initially holds different expectations about the future of asset prices. There are no publicly known expectations: each agent forms his or her expectations on the basis of their anticipation of other agents' expectations. Observing the market, they look for profitable opportunities to trade. As time goes by, they individually form hypothetical expectational rules, test these, trade on the ones that predict best, drop those that perform badly, and introduce new ones to test. Individuals' expectations therefore evolve and 'compete' in a market formed by the expectations of others. In other words, agents' expectations 'coevolve in a world they cocreate'.[10] Prices of assets are driven endogenously by these induced expectations.

Using computer simulations of this artificial stockmarket, the authors were able to explore alternative outcomes arising from varying selected assumptions. It turned out that when traders were allowed to respond rapidly to new market observations, the market self-organised into a 'rich' set of expectations. Temporary bubbles and crashes emerged, while trading volumes were high, with periods of quiescence alternating with periods of intense activity. The price time series showed persistence in volatility, another characteristic observed in actual markets; and individual behaviour evolved continually and did not settle down.

Only when agents were allowed to adapt their expectations to new observations of the market's behaviour at an unrealistically slow speed, did the market converge to a rational-expectations regime. In this regime, none of the observed characteristics of actual financial markets were manifested.

What is significant about this work is that it retained most of the other assumptions of traditional neoclassical equilibrium theory. Each individual was assumed to optimise their portfolio, and full market clearing was achieved in each time period. Only in the way in which expectations were assumed to be formed did it depart from the conventional approach. The remarkable difference in the results obtained underlines the inadequacy of the rational expectations hypothesis.

Perhaps the strangest legacy of rational expectations theory is that, when combined with a belief in monetarism, it produces the conclusion that all that a government need do to achieve a lower rate of inflation is to announce a target for the growth of the money supply. Once people are convinced of the determination of that government to stick to its declared policy, they will form their expectations of the likely future rate of inflation and adjust their behaviour accordingly. This proposition requires the simultaneous fulfilment of all of a number of heroic assumptions. These include the assumptions that the money supply can be measured, that the government is in full control of it, that there is a predictable relationship between the money supply and the rate of inflation, and that any democratically elected government will sustain an unwavering commitment to anything for very long. Any doubts on the last point, so far as the UK was concerned, were resolved by the suspension of sterling from the ERM in September 1992, a policy to which both government and opposition had repeatedly asserted there was no alternative.

The theory of economic growth

Of all the branches of economic theory, economic growth has always proved to be the most fascinating. It is so because every politician would like to know the formula for a society to progress from poverty to riches. As a matter of fact, Adam Smith hinted at just such a formula when he wrote that 'Little else is requisite to carry a state to the highest degree of opulence from the lowest barbarism but Peace, Easy Taxes, and a tolerable administration of Justice; all the rest being brought about by the natural course of things'.[11] Over the past 200 years a growing number of countries have moved to a state of opulence, if not from barbarism then at least from a state in which great poverty seemed the fate of the majority of the people.

Economic historians are not agreed on a simple explanation of how this came about (but see Chapter 10). What is commonly recognised is that even a list of the proximate determinants of growth would be a long one, with complex interactions between them. Furthermore, these determinants themselves are the result of the interaction of factors which lie primarily outside the boundary of economics, in the realms of psychology, sociology and politics.[12]

These considerations have not deterred equilibrium theorists from extending their model-building efforts into the area of economic growth. They have done so simply by ignoring all 'non-economic' factors, such as institutions and motives, as well as those economic factors which are not easily quantifiable, such as entrepreneurship, confining themselves instead to measuring the relationships between a handful of aggregates like GDP, total investment, total capital stock, the rate of population growth and the like.

The extension of equilibrium economics into the area of economic growth is a postwar phenomenon, and has proceeded in three waves. The first wave was founded on the work of Harrod and Domar, who in turn had built on the earlier work of Cassel. The fundamental idea of the Harrod–Domar models was that for each national economy there should exist a unique rate of growth which would just bring the annual addition to its aggregate productive capacity into balance with the annual increment in aggregate demand. This was known as the balanced growth rate or the equilibrium growth rate. When Abramovitz conducted a survey of the state of economic growth theory for the American Economic Association (AEA) in 1952 he left these equilibrium growth models out of his consideration on the grounds that 'they offer no explanation or prediction of the way in which actual economies have developed or are likely to develop over time'.[13]

A second wave, one might almost say a tidal wave, of writing about equilibrium growth models, was triggered by the introduction by Solow in 1956 of the so-called 'neoclassical aggregate production function' into the Harrod–Domar model. This greatly increased the flexibility of the equilibrium growth model, and the flood of literature released by this development swept away any resistance on the part of the AEA's new reviewers, Hahn and Matthews, when they came to update Abramovitz's survey in 1965. This time the review of growth theory was devoted exclusively to equilibrium growth models. However, the reviewers were bold enough to point out in a footnote that the theory they were surveying 'was different from that which would be used if the immediate purpose was to provide the best available explanation of the variety of historical growth experience', adding rather lamely that 'the authors of these models have naturally had in mind as a rule that their work should contribute to an understanding of the way economies actually grow over time'.[14]

So in the area of economic growth as in other branches of

equilibrium theory, progress was not directed to a better under-standing of observed phenomena but to the construction of a mathematically more sophisticated theory. Once again the el-egance of the scaffolding took precedence over its purpose and usefulness. According to equilibrium growth models, all capital and labour was fully employed at every moment in time, and the whole capital stock grew at the same steady rate through time as did the labour force. Every industry grew at the same rate, and the composition of output did not change. There were no new products, no new processes, no new mineral deposits were discovered, none were exhausted, no new markets opened, none disappeared. There were no taxes, no governments, no tar-iffs, no quotas, no transport costs. It was a model in which the whole economy glided smoothly through time, like a skater on ice – a moving equilibrium with every part in perfect harmony. This was called 'steady-state' growth.

In steady-state growth models only one factor determined the rate of growth of the model economy, and that was the rate at which aggregate productivity increased. Progress in technology leading to increases in aggregate productivity took place in some unexplained way outside the model. A further unsatisfactory feature of these models, a consequence of their mathematical proper-ties, was that as capital accumulated through time, the rate of growth of income per head would eventually fall to zero, an outcome which seemed to contradict historical experience.

Just when economists had begun to tire of playing with steady-state models, equilibrium growth theory was given a new lease of life in the late 1980s. The system of mathematical equations extended by Solow was further extended by Romer[15] so that the causes and effects of technological progress could be incorpor-ated within it, rather than being determined outside it. Thus was launched the third wave of equilibrium growth theory, un-der the grandiloquent title of 'post-neo-classical endogenous growth theory'.

In plain English, this new extension was saying no more than that investment in commercial research and development leads to the discovery and adoption of new technologies and new products which in turn increase productivity. While many people would regard this as a fairly commonplace observation, whose practical importance had been emphasised by Schumpeter some 50 years earlier, it was hailed by its proponents as a great 'theoretical advance'. Needless to say, all the important subtleties of the social and business relationships involved in the acquisition, transmission and implementation of new knowledge which had been identified by earlier generations of economists were left out of the model. Instead, these complex relationships were reduced to a rather crude mechanical relationship, in which one variable representing the aggregate rate of increase in productivity was asserted to be proportional to some measure of 'new knowledge'. But this was enough to give economists the excuse for experimenting with the altered model, rather than looking more closely at the actual experience of economic growth in the real world.

What excited equilibrium theorists about the new model, apart from the fact that the rate of technical progress was determined within the model (hence *endogenous* growth), was that it appeared to dispense with two traditional assumptions which were clearly at odds with reality, namely perfect competition and constant returns to scale. What was overlooked was that in order to maintain precisely determinate outcomes, the new model introduced two even more unrealistic assumptions, constant returns to capital and 'infinite-horizon intertemporal optimisation'. The latter simply extends to the indefinite future the assumption of perfect knowledge made by the conventional static equilibrium model. Each firm and household is assumed to foresee all future prices correctly to eternity. The former is a knife-edge assumption. With anything less than exactly constant returns to capital, the model cannot produce sustained growth. With anything more, the model explodes. As Solow commented, anyone who believes that the real world is characterised by exactly constant returns to capital also believes in the tooth fairy.

The endogenous growth model retains all the other limitations of equilibrium growth models, including the omission of the changing composition of aggregates. As economies grow, the sectoral composition of output typically shifts in emphasis from agriculture to manufacturing then services. Even more important is the omission of the consideration of any human factors. As Abramovitz has observed:

> ... current investigations of underdeveloped areas have pro-
> duced no more universal finding than that the process of
> investment is frustrated not by any lack of potential opportu-
> nity for gain and not by any lack of surplus saveable income,
> but rather by the lack of a vigorous business class, intent on
> searching out and exploiting local business opportunities.[16]

What seems particularly inappropriate in applying equilibrium
methods to the process of economic growth is that, being a long-
term process, it is characterised by cumulative rather than repetitive
change to an even greater degree than other economic phenomena. Thus the proper methods of study should be closer to history
than to mechanics. In the case of the endogenous growth model,
the reduction of the complex interactions of factors involved in
the process by which investment in research and development
is turned into improvements in productivity, with all the variable time lags and uncertainties, to a single mechanical relationship
between two aggregate variables might be acceptable as a back-
of-an-envelope calculation but can hardly be regarded as a
contribution to scholarship. It has added nothing to our understanding of how economies actually grow over time. The huge
amounts of time and effort which have gone into constructing
equilibrium growth models have been diverted from the work of
looking for regularities in patterns of historical change. Instead,
equilibrium theorists have again preferred the easier task of discovering the logically necessary implications of arbitrarily chosen
premises.

Needless to say, the endogenous growth model has not been
subject to direct empirical tests. But, just when equilibrium theorists
had concluded that productivity increases were the result of ex-
ternalities and the production of new knowledge (these results
being implicit in their assumptions), there was emerging amongst
those organisations whose function it is to promote growth in
developing economies a rediscovery of Adam Smith's perception
that the most important ingredients in any formula for economic
growth are institutional. And so endogenous growth theory has
proved to be yet another ephemeral intellectual fad.[17]

Econometrics

One other postwar development epitomised the pseudo-scientific characteristics of equilibrium theory. This was the topic, or perhaps one should say the practice, called Econometrics, which rapidly grew from the 1950s to become a field of study in its own right. Even today, academic posts are to be found in some universities with titles such as Lecturer in Econometrics, although if the limitations of the technique were more widely understood, it is doubtful they would still survive.

Econometrics starts with the usual equilibrium model made up of a set of mathematical equations that we encountered earlier. As an additional gesture towards scientific method, the parameters of these equations are then estimated by applying statistical procedures to 'fit' the equations to some observed data. In the early days of econometrics, in the 1950s and 1960s, it was expected that such econometric models of the whole economy could successfully be used for forecasting purposes, while the method itself could also be used to test the validity of particular economic theories. Both of these hopes proved to be false.

First, as we shall see, the nature of economic activity in an advanced market economy is such that precise predictions of particular events are simply not possible. Mechanical theories which pretend otherwise must therefore produce answers which will only be correct by chance. The second hope foundered on the difficulty that economic theories had to be seriously distorted to squeeze them into the straitjacket of statistical testing procedure. Experience showed that, as Richard Layard observed, those econometric models which did in a statistical sense fit the facts had no theoretical basis, while those that were founded on economic theory (however tenuously) did not fit the facts.

Moreover, the data on which the parameters of econometric models are based are drawn largely from official statistics. Anyone who looks through an official statistical publication will find that the data in most of the important series are subject to later revision. Sometimes, the size of these revisions can be half as much again as the change initially reported. An increase in the value of a particular statistic can even be turned by a subsequent revision into a fall. And the revisions cease, not so much because the statistician has finally arrived at a perfectly accurate estimate, but more probably because (s)he is required for other, more pressing, tasks.

The British government does not even publish a prominent health warning alongside each number, as the US government does. For example, the US Census Bureau may say that Housing Starts rose in a given period by 1.5 per cent plus or minus 2.4 per cent. The reader therefore knows it is not even certain that housing starts rose at all. In the UK, one has to read the small print. But the econometricians go on estimating their parameters regardless of these uncertainties. Even if they had been based on perfect information, such parameters would have been, in any case, of doubtful significance. The theoretical requirements for an acceptable procedure for statistical estimation of the parameters concerned are hardly ever satisfied by official statistics.[18]

Conclusions

The development of equilibrium economics since Marshall has been a story not of intellectual progress, but of intellectual retreat. We are now further away from, not closer to, an understanding of how a modern economy actually works. The assumptions of the most recent equilibrium theories are far more demanding than the assumptions of the scheme of general equilibrium originally put forward by Walras more than 100 years ago. Like Keynes, he would most probably have strenuously objected to the developments subsequently carried out in his name.

The failure of equilibrium economics has not escaped the attention of students. On both sides of the Atlantic, they have been deserting the subject in droves. In the UK in 1995 the number of school students enrolled for 'A-level' Economics was one-third down on 1990. Young people find what is now taught in the Economics curriculum to be boring and unrelated to their experience of the world. Many drift into business studies, where the curriculum is in any case less demanding. The widening gulf between equilibrium theory and the reality of the market economy has presented an opportunity for sociologists, business school teachers and even management consultants to put forward ideas of their own, and some of these are reviewed in Chapter 5.

But, as we shall show in the next chapter, the most hurtful blow to the pretensions of equilibrium economists has come not from business but from governments. Because of the failure of equilibrium theory to provide a reliable intellectual framework for government policy, policy-making has finally detached itself from that theory in the two areas where it had most hoped to make a decisive contribution, namely in monetary and fiscal policy. The lesson for the economics profession is that it must abandon its pretensions to a knowledge it does not possess. Our present understanding of economic activity does not provide a basis for economists to act as if the economy were a machine and they were engineers. If we are to improve our understanding of how the economy actually works, it will come about through adopting a very different approach to the one that has dominated economic theory in the last fifty years.

4
Economics and Politics

'Except on matters of detail, there are perhaps no practical questions, even among those which approach nearest
to the character of purely economic questions, which
admit to being decided on economic premises alone.'
J.S. Mill, *Principles of Political Economy*

Since the end of the Second World War, the state of the economy
has nearly always been the dominant factor in influencing the
outcome of elections in the Western democracies. It has therefore been essential for politicians to convince their electorates
that they are in control of economic events. After the war, economists were happy to assure the politicians that, thanks to the
new 'scientific' equilibrium economics, they really would be in
control. With their hands on the levers of economic policy, the
economy would respond like a machine. The politicians and the
voters believed them.

It was true that in the immediate postwar period in Europe,
governments really could control their economies simply by keeping in place the emergency powers which they had acquired in
wartime to control prices and to allocate goods and labour by
decree. The controlled economy worked very much like a Soviet-
style planned economy. Gradually, wartime direct controls were
withdrawn and markets de-regulated, but economists believed that
governments could still achieve the principal objectives of postwar economic policy. These were:

1. Full employment;
2. Sustained growth of total output; and
3. Price stability.

In the early years after the war, the overriding concern throughout the Western world was to avoid a repetition of the levels of unemployment which had been seen in the 1930s. The Great Depression had had a particular devastating effect on the United States. By the end of the war, in 1945, the economics profession had almost unanimously accepted the analysis of the unemployment problem which had been advanced by John Maynard Keynes in his 1936 book, *The General Theory of Employment, Interest and Money.*

The main practical implication of Keynes' new theory was that economic depressions and their associated unemployment were avoidable. They were the result of an insufficiency of total spending within an economy, which it was within the power of governments to correct by spending more money at the right time and in the right amounts. Whereas the existence of unemployment indicated an insufficiency of total spending in a national economy, the existence of inflation indicated too much. All that governments had to do, according to the Keynesian version of equilibrium theory, was to make sure that the level of spending was neither too little nor too much, but just right, so that the three objectives of full employment, sustained growth and low inflation could all be achieved at the same time. And economists, using their mechanical models of the economy, could not only work out how much spending was needed, but could show the politicians how and when to do it, by raising or lowering tax rates or by altering the level of government spending. That was known as macroeconomic management or demand management.

At first, the policy of demand management appeared to work, at least so far as the principal objective of full employment was concerned. For some thirty years after 1945, unemployment rates throughout the Western world remained at historically low levels, and downturns in economic activity were so mild and infrequent that they came to be known as recessions rather than depressions. The relative mildness of these recessions in the early postwar period can be seen with hindsight to have been due in large part to the increased share of the public sector in postwar economic activity.

The first decade after the war was marked in many European countries by the nationalisation of large parts of the economy, notably the railways, gas, electricity, telecommunications and some of the heavy industries such as coal or steel. At the same time, there was an increase in rates and in the coverage of taxation to help governments pay for expanded health, education and other welfare services as well as social security benefits. One of the results of these developments was that whenever an economy began to move into recession, tax revenues would begin to fall while unemployment benefit payments rose. By automatically stabilising in this way the total amount of spending in the economy, the severity of depressions was alleviated. Conversely, the consequences of temporary excesses of expenditure were dampened by the fact that tax revenues rose sharply in a boom, while more people being drawn into work reduced the amount of unemployment benefit being paid out.

On the other hand, active government intervention as envisaged by equilibrium theorists in the shape of counter-cyclical monetary and fiscal policies proved to be ineffective in practice – it even made things worse – because of the problem of time lags between the perception of the need for such measures and their effect. In order to avoid the destabilising effects of increasing total spending after a recession is over, expansionary measures needed to be implemented *well before* a cyclical upturn. But, as we saw in the last chapter, turning points in the business cycle are precisely what economists have not been able to predict.

The apparent disregard which Keynes had shown for the possible dangers of inflation appeared to be justified by the events of the 1950s and 1960s. In Britain, the rate of inflation averaged less than 4 per cent in those two decades, and when it rose above 5 per cent in 1969 it was the first time for 14 years. Whenever inflation did threaten, nationwide collective bargaining by trade unions was held to be responsible. As a result, wage freezes were periodically imposed by governments, dignified by the name of 'incomes policies'.

This happy period, now looked back on by many economists as a Golden Age, was brought to an abrupt end by the explosion of world oil prices in 1973 and 1974. The wholesale price index for crude oil in the UK rose from 22.5 in 1972 to 149.3 in 1977

(1975 = 100). The oil price increase triggered an inflation of prices throughout most of the Western world, and at the same time brought about a slump in output and a rise in unemployment. These simultaneous developments discredited the then conventional wisdom, the Keynesian theory, which held that unemployment and inflation could not exist at the same time. That view was publicly repudiated in 1976 by the then British Prime Minister, Mr James (now Lord) Callaghan at the Labour Party Conference of that year, when he warned his audience that a government could not spend its way out of a recession.

But a new orthodoxy was already emerging to take the place of the old. It was called Monetarism. Monetarism had begun to enter British government thinking in the mid-1970s, and was put into practice enthusiastically by the new Conservative government which came to power in 1979. It was also adopted by the Reagan administration which took office in the United States in 1980. Monetarism reversed the priorities and instruments of earlier economic policy. Hitherto, the principal objective of government policy had been to secure full employment, while the control of inflation was more or less left to market forces. Under monetarism, the principal objective of government policy became the control of inflation, with the realisation of full employment left to market forces. Unemployment was no longer thought to be the result of insufficient total spending, but rather of inflexible labour markets. In other words, those things which discouraged companies from offering employment like minimum wages, taxes on employment, employment-protection legislation and the unreasonable demands of trade unions, were now believed to be the true causes of the problem of unemployment, rather than inadequate aggregate demand as had previously been supposed.

Under the old Keynesian conventional wisdom, a government could choose between having more unemployment or more inflation, according to how little or how much it spent. Under the new monetarist doctrine, there was no such trade-off. The achievement of price stability, that is the control of inflation, was held to be a prerequisite of the sustained growth of output and employment. In this view, inflation was caused not by excessive wage demands, as many people had earlier argued, but simply

by too much money circulating in an economy. Inflation was therefore capable of being restrained through the adoption of appropriate
policies by governments to control the supply of money.

Although the policy shift from Keynesianism to monetarism
in the UK can be traced to 1976, supporters of the Keynesian
view did not give up easily. For many years after 1976 the British public were first amused and then bored by the acrimonious
sectarian disputes between the two camps. This brawl did great
damage to the reputation of the economics profession until, in
the end, the monetarists achieved a great victory. The decisive
blow against the Keynesians fell in March 1981, when 364 of
their number, including all their principal supporters, signed a
letter to *The Times* predicting that Chancellor Geoffrey Howe's
monetarist policies would prevent the economy recovering from
that year's recession. The economy recovered.

Despite their differences of opinion, neither of the two con-
tending groups were in any doubt that a government, if it so
wished, *could* control the economy in such a way as to realise
its policy objectives. Indeed, the theoretical models of the economy
employed by the two groups were remarkably similar: both shared
a mechanical view of how a market economy worked. The only
difference between them was that monetarists believed in the
existence of a simple and stable relationship between the supply
of money and the level of prices, while the Keynesians believed
in an equally straightforward relationship between the level of
aggregate spending and the level of total employment.

Mrs Thatcher was elected in May 1979. After a brief but deep
recession (interpreted by monetarists as a painful but necessary
adjustment), the British economy grew quite rapidly for some nine
years from March 1981. By 1986, inflation seemed to have been
conquered, and if full employment had not yet been realised,
unemployment had been falling for three successive years. Once
again, the politicians and their economic advisers allowed themselves to believe that they were in control of events. Once again,
these illusions were shattered. Following a sharp fall in the stockmarkets in both London and New York in October 1987, monetary
policy was eased on both sides of the Atlantic for fear of a possible
recession. There followed instead an unexpected inflationary boom

from 1988 to 1990, succeeded by a recession from 1990 until 1992. This last period was marked by periodic and confident predictions by City and Treasury economists alike of imminent upturns which never materialised.

By this time, the patience of government policy-makers with their economic advisers was beginning to wear thin. The failure to anticipate either the inflation of 1988–90 or the timing of the recovery from the recession of 1990–92 caused a reexamination of the postwar forecasting record of government economists in the Western world. The survey confirmed that turning points had never been predicted, and that even in normal years the official forecasts nearly always turned out to be misleading. Nor had these forecasts improved over the years, no matter how many more equations had been added to the models and no matter how many more economists had been employed to tend them. It was clear that governments and their economic advisers were not in control of events. Nor was their understanding of the economy good enough to allow them to be able to anticipate what might happen with any degree of accuracy. So far as the UK was concerned, all remaining credibility was lost on Black Wednesday, 16 September 1992, when speculation against the pound in the foreign exchange market forced the Treasury to abandon its target exchange rate of DM2.95 to the pound sterling. This was the rate the government had insisted upon two years earlier, against the advice of their partners, when bringing sterling into the ERM. In a futile attempt to resist the tide of market forces and 'save' the exchange rate, the Treasury lost, and speculators gained, some £10 billion of British taxpayers' money in one afternoon. The policy of fixing the exchange rate of the pound to the German mark whatever the domestic cost, a policy to which both government and opposition had agreed there was no alternative, became redundant overnight.

Following the departure of sterling from the ERM in September 1992, the Conservative government adopted a policy of targeting the rate of inflation directly. Official short-term interest rates were to be the principal instrument to achieve the stated objective of an annual rate of 2.5 per cent or less, and explicit targets for 'intermediate' money supply variables were dropped. This policy has remained broadly unchanged ever since, although the incoming Labour government of 1997 gave the Bank of England operational independence, and made the 2.5 per cent inflation target symmetric.

It seems curious to suppose that having abandoned earlier attempts at targeting intermediate variables such as the money supply or the exchange rate, the monetary authorities could be any more successful at achieving an inflation-rate target. It is even more curious when it is realised that what is being targeted is not the current, but a forecast rate of inflation. In other words, the principal instrument of monetary policy, short-term interest rates, are altered not in response to current movements in the price level, but in response to the Bank's forecast of what it will be in two years' time, that being the supposed time lag. Given the consistent failure of official attempts at forecasting the rate of inflation (see Chapter 3), this seems to be an experiment which owes more to hope than to experience.[1]

Although the inflation-rate target has been broadly achieved since 1992, this has perhaps owed as much to good fortune as good judgment. Over the period concerned, the UK's leading trading partners have also been pursuing low-inflation policies, and the world has generally been free of inflationary shocks or surprises. What is clear is that insofar as the success of monetary policy in the UK since 1992 has been based on domestic monetary policy decisions, these decisions have been based on pragmatic considerations and not on macroeconomic theory. Those concerned with the actual operation of monetary policy during the period have admitted that there is an absence of consensus about the way in which interest rates affect inflation. The channels of influence are complex, involving the exchange rate, the housing market and other asset prices. And even if these linkages were understood in principle, it remains widely recognised that the time lags involved are long and variable.[2]

What this means is that in the UK the conduct of monetary policy has become uncoupled from macroeconomic equilibrium theory. The same is apparently true in the United States, where the Federal Reserve Board is reported to have abandoned its traditional analytical tools, recognising that they are no longer appropriate. Speaking of the formal basis of monetary policy decision-making, Alice Rivlin, Vice-Chair of the Fed, is reported as saying: 'I don't think we really have a lot of clues. The macro-statistics may not be all that much help. People who want to understand this may need to pay a lot more attention to what is really going on in businesses and labour markets.'[3] If there is one positive lesson to be learned from the constant changes of monetary policy regime in the UK since the war, it is that, like other participants in the human economic system, governments proceed by trial and error.

Fiscal policy in the UK

Up to the late 1940s, the rhetoric if not the real influences on public expenditure and taxation in the UK reflected the traditional concerns of sound public finances and balancing the budget. In the 1950s and 1960s there was a gradual acceptance amongst government policy-makers of the Keynesian idea that the public finances should be used to manage the level of aggregate demand, so as to achieve the objectives of economic policy, of which the overriding one was the achievement of full employment. No sooner had this Keynesian idea become the conventional orthodoxy than it was challenged by monetarism. During the Keynesian–Monetarist debates of the 1970s and early 1980s, both sides appealed to macro-economic theory to support their respective positions. As the debates died down in the mid-1980s, the rules of fiscal policy gradually became disconnected from the objectives of macro-economic policy, full employment or the control of inflation. In the Budget of 1988, the then Chancellor restored the principle of a budget balanced over the business cycle, justifying this rule with reference to the burden of debt service and future levels of taxation. This rule remained in force, at least in principle, until the Labour government which came to power in 1997 replaced it with two new budgetary rules. The first, a so-called 'golden rule', required government current expenditure to be covered fully from taxation. The second, christened the 'sustainable investment' rule, limited net public debt to a 'stable and prudent level', set initially at 40 per cent of GDP on average over the business cycle.

Neither of these two rules, nor the purely arbitrary number of 40 per cent, have any basis in macroeconomic theory. Indeed, the Treasury paper in which they were presented, 'Stability and Investment for the Long Term', made no mention of a possible link between fiscal policy and employment, and said nothing about any measure of the money supply. The rationale of the new rules expressed in terms of 'stability' and 'prudence' and the matching of new debt with productive assets indicate a return to prewar, even Victorian, budgetary concepts. This may not be accidental. As Congdon[4] points out, in terms of results, fiscal policy in the UK was more successful before the 1960s and after the mid-1980s. It was in that unhappy intervening period that macroeconomic theory and analysis had some influence on British fiscal policy.

The liberalisation of the market economy

For one hundred and fifty years up to the 1980s governments in the more advanced industrialised economies intervened on an increasing scale in the operations of their market economies. They did that by three principal means – by direct regulation of various aspects of economic activity, by increasing the level and extending the scope of taxation, and through the extension of public ownership of industry. This tendency towards increased intervention accelerated slightly after the First World War and most markedly after the Second World War.

Although the speed and extent to which this happened was more pronounced in some countries (Sweden and the United Kingdom) than in others, (Switzerland and the United States), the tendency had been in existence for so long that it seemed to acquire a historical momentum which many thought would lead inexorably towards complete socialism. Of course, some of those who believed this hoped that it would happen. But the apparently inexorable process whereby capitalism was being gradually transformed by increasing government intervention into socialism faltered for the first time in the 1970s, and then went into reverse in the 1980s.

The 1980s will be remembered by historians as the decade that marked the beginning of a completely new era in the history of the relationship between governments and market economies in the Western world. Governments began to dismantle their instruments of economic control by deregulating industries, lowering rates of taxation and turning over state-owned industries to private ownership. Quite suddenly, instruments which had hitherto been seen as an essential part of a government's mechanism for control of a modern market economy were now seen as among the main factors inhibiting its successful performance.

Beginning in the United Kingdom, the United States, and perhaps most dramatically in New Zealand, a tidal wave of deregulation and privatisation was unleashed which subsequently spread throughout the rest of the world. This gave the market economy a scope it had not enjoyed since before the First World War. The abrupt and unexpected collapse of the Soviet Union at the end of the same decade, brought about by the failure of its planned economy, appeared to underline the message that the fewer restrictions are imposed on a market economy, the better it will work. This is a lesson which countries have been slow or fast to learn depending upon the urgency with which they felt the need to increase their material standards of living. Some poorer countries have been quick to embrace the new message, while some older Western European countries have been more reluctant to change their ways. Of course, different countries have started from very different positions. At one end of the spectrum were the former communist countries of Eastern Europe with planned economies, and those Third World countries who, in the early postwar period, believed that central planning was the fastest route to raising their living standards. At the other end were countries like Switzerland and the United States where at the beginning of the 1980s market forces already played a dominant role.

However, it was in the USA and UK that the process of liberalisation may be said to have begun. In the UK, Mrs Thatcher began by removing controls on capital movements and on foreign trade, and then lowered the higher rates of tax on income, increasing taxes on consumption. At the same time attempts were made to lighten the burden of regulation on individual industries – with mixed results. She stumbled almost accidentally into the privatisation of state-owned industries: it had not featured in her 1979 General Election manifesto. In the USA there was no scope for privatisation, since the Federal government did not own any industries. But President Reagan cut income tax rates in the face of a large existing budget deficit, and launched a campaign against Big Government, that is, he attempted to cut back on Federal government non-defence spending.

To see more clearly what this process of liberalisation really meant, we can look at the experience of New Zealand.[5]

The New Zealand experience

New Zealand was one of the first countries in the world to provide a cradle-to-grave welfare state, something it had already achieved before the Second World War. By the early 1980s it had become one of the most regulated and protected societies in the Western world, and the result was a sluggish economic performance. Between 1950 and 1984, the New Zealand economy had grown at only half the average rate of the other OECD member states.

After a foreign exchange crisis on 15 July 1984, when the Central Bank was forced to suspend all trading in foreign exchange, the newly elected Labour government realised that a drastic reversal of policy was required. It embarked on a programme which over the next twelve years transformed New Zealand from being one of the most regulated economies in the OECD to being among the least regulated of all. These reforms included the removal of controls on wages, prices and foreign exchange, and the flotation of the New Zealand dollar; import quotas were removed and tariffs reduced; and subsidies to agriculture and industry virtually disappeared. The most remarkable liberalisation took place in the labour market. Since 1991, contracts of employment have been on almost the same basis as any other commercial contract. By December 1995, only 17 per cent of the workforce had union-negotiated collective contracts. High marginal rates of income tax were reduced, and a broadly based value added tax introduced. In addition, many state-owned companies have been privatised. Since 1989, the government has set a target rate of inflation, and left the Central Bank to implement it.

The results of lifting the burdens of regulation, taxation and public ownership from the market economy in New Zealand were impressive. Unemployment fell from a peak of 11 per cent in 1991 to 6.2 per cent in 1995. Inflation, which had been over 15 per cent in 1985, remained below 2 per cent between 1990 and 1995. Investment is at historically high levels, and a sustainable growth rate of total output of between 3.0 and 3.5 per cent per annum is believed to be possible, higher than that of which most other mature OECD economies seem to be capable.

Political responses to the economic process

The spread of the process of abolishing controls across the world has meant that many of the barriers which governments had erected over time to the free movement of goods, services and money between countries have been dismantled. Individual countries have seen for themselves the benefits of doing so. Together with a reduction in worldwide transport costs and journey times, there has developed for many commodities what is effectively a single world market. Nowhere is this more evident than for those goods (clothes, shoes, drinks) and services (music, videos and computer games) which are the tokens of popular culture. This is part of the process known as globalisation.

Globalisation is already having repercussions in the Western countries where it began, and not all of these repercussions are favourable. While citizens of the West like buying cheap clothes and other manufactured goods imported from poorer countries and enjoy taking holidays in these countries, they are not happy with the job insecurity which freer trade with an ever-widening range of countries has brought. The growth of protectionist sentiment in the United States has generated continuing trade disputes with Japan and the European Union, and widespread hostility to further measures for trade liberalisation. In Europe, unemployment has risen to unprecedented levels in some of the major economies. There are many politicians who believe it would be a good idea to build protective barriers on the external frontiers of the European Union.

It is true that the lifting of barriers to trade is not the only threat to jobs in a modern economy. The accelerating rate at which new technology is introduced is an even more important destroyer of jobs. But, contrary to what many people instinctively believe, neither of these reduces the total number of jobs available. In fact, the opposite is true – they both create more jobs than they destroy. As a result of the last great wave of global trade and new technology which hit Great Britain in the nineteenth century, the population grew from 14 million in 1821 to 33 million in 1891.

But it is also true that the changes brought about by free trade and by the introduction of new technologies are highly disruptive to normal life. The new jobs which are created may be in different industries, may require different skills, may be located in other parts of the country, and may be lower paid than the old jobs. This disruption bears harder on some people than others. It has been harder for those who for one reason or another are less mobile, those whose skills are less versatile than others, those who, in other words, are less adaptable than others. It also bears down disproportionately on the owners of those types of capital that are specific to one particular use. Equipment which has been built to produce vinyl records cannot be easily adapted to producing CDs, and therefore its value can quickly be destroyed. Disruption is politically unpopular, and the greater the disruption the stronger are the popular political pressures on government to modify the working of market forces. Resistance to the disruptive effects of market forces in the Western world has most importantly taken the form of attempts to moderate their influence in the labour market. This has been one of the functions of trade unions, and in many European countries the activities of trade unions have been given special legal protection. Governments have also legislated in favour of minimum wages, and to restrict the ability of employers to dismiss workers unconditionally. However, the experience of the past 50 years tends to suggest that the net effect of such restrictions is to slow down the adoption of the innovative working practices which contribute to higher productivity per worker. Together with heavy social security contributions which employers must pay for each worker they hire (a tax on jobs), this has discouraged employers from taking on workers.

For example, in the USA, where the hiring and firing of labour is much less hampered by law and by trade union activity than in Western Europe, total employment increased at an average rate of 2 per cent per annum between 1960 and 1988. In the same period, in Western Europe (OECD Europe) the total number of jobs increased by only 0.4 per cent per annum. In the decade from 1988 to 1998, US employment continued to expand at an average rate of 1.6 per cent, while the EU could achieve only 0.4 per cent. It is generally recognised that the principal reason for these differences is the greater flexibility of labour markets in the USA; that is, workers there are much more willing to change their jobs, their terms and conditions of work (including reductions in pay), and to move to other parts of the country when necessary. The same difference in flexibility probably explains much of the difference in rates of unemployment between continental Europe on the one hand, and the USA on the other. For example, the unemployment rate in the USA has fallen from over 7 per cent in the early 1980s to less than 5 per cent today. In France it has gone from 7 per cent to around 12 per cent over the same period. In Germany unemployment has risen from 4 per cent in 1981 to over 9 per cent in 1998. Given the high levels of unemployment in these two countries it is not surprising that their policies should be following the example of the UK which began to take steps in the 1980s to deregulate its labour market.

While the removal of many traditional restrictions on the free movement of goods, services, capital and labour, not only between countries but between occupations has led, via the intermediation of competition and innovation, to a spectacular rise in material living standards throughout the West, it has also had other consequences. The degradation of the physical environment has given rise to political demands for new kinds of restrictions on market forces. The disruptive effects of continuing and unpredictable changes in the social environment have led to increasing reports of stress and lack of satisfaction in work, and an apparently growing dissatisfaction with what is called the quality of life. There is concern in almost every Western country about the apparently unstoppable increase in rates of crime and other indications of social disorder such as suicide, drug-taking, alcoholism, and so on. It is difficult to believe that these symptoms are not related directly or indirectly to the disruptive effects of market forces.

In the United States, the United Kingdom and to a lesser extent in continental Europe, someone who is a political conservative normally supports the principle of free competition; that is, the free play of market forces with minimum government intervention. Conservatives recognise that competition is the means by which the benefits of foreign trade and of innovation (new goods and new technologies) are spread throughout a country. Yet at the same time it is the disruptive effects of changes brought about by the forces of market competition which undermine other institutions which conservatives hold dear – notably traditional attitudes and values and the organisations which embody them.

So here is a paradox. The very means by which the burden of poverty – the major source of human unhappiness throughout history – has been lifted – namely the unfettered operation of market forces – is at the same time the apparent cause of many new sources of human unhappiness.

Given the diminishing urgency of an individual's or a society's material needs as they grow richer, one solution might be to slow down the rate of change – and thus the rate of disruption brought about by market forces. But to believe this is to repeat the mistake of thinking of the economy as a machine, whose speed can be controlled with predictable results.

Conclusions

If market forces are to be the servant of a society rather than its master, then they must evidently be controlled. But in order for control to bring about the desired results there must first be an understanding of how the economy actually works.

The economic policy of governments in the Western world in the postwar period have been based largely on the perception of the economy as a machine. It was believed that by adjusting the levers of control, principally fiscal and monetary policies, governments could obtain the results they desired – full employment, sustained growth and stable prices. But policies based on equilibrium theories did not work, and the experience has led governments to uncouple their policies from such theories. The equilibrium models have not been able to explain the great surge in productivity in the United States between 1950 and 1975, and the slowdown thereafter. Nor are the equilibrium models able to explain the achievements of successive rounds of international trade liberalisation conducted under the auspices of GATT (now WTO), nor the success of liberalisation within the market economies brought about by deregulation and privatisation, in contributing to the growth of output and employment. Nor can equilibrium models explain why prolonged booms in market economy countries should be punctuated by periodic busts.

Outside the advanced economies, the performance of economic policies led by equilibrium theory has not performed any better. In Russia, Western policy makers, whatever their individual differences of emphasis, have urged 'macroeconomic stability' as the principal objective of economic policy to the exclusion of the more important institutional considerations, with tragic consequences. Likewise, in the developing economies the importance of institutional factors has until quite recently been downplayed, presumably because such factors did not feature in equilibrium models of economic growth.

In all of these cases, successful policy depends on a tolerable understanding of how market economies work. The mechanical paradigm of the equilibrium theory does not provide this. In Chapters 7 and 8 we shall introduce a theory which offers the prospect of a better understanding. Meanwhile, in the next chapter we look at the ways in which equilibrium theories have impinged on thinking about that part of economic activity which takes place within business organisations.

5

Economics and Business

'... the Learned give up the evidence of their senses to
preserve the coherence of the ideas of their imagination.'
Adam Smith, *Essay on the History of Astronomy*

Up to the Second World War, when economists still wrote their
papers in plain English, it was possible for businessmen and economists to communicate with one another and thus to learn from
each other. They shared a common vocabulary and a common
interest in the same phenomena: how markets work, productivity and profitability, and the relation between costs and prices.
It is true they were looking at these phenomena from different
perspectives. The businessman looked at economic activity from
the point of view of his firm and his industry, while the economist took a wider perspective of the interaction of the whole
system of firms, industries, consumers and government. Economists looked at the performance of the economy as a whole.

After the war, businessmen discovered that the new 'scientific'
equilibrium economics had no place for those things, like enterprise, innovation and profit, which they believed to be rather
important. Economists for their part were impatient with
businessmen's apparent lack of interest in their new models and
their new vocabulary, and were accordingly dismissive of business
views on questions of economic policy. Still, the new economics
claimed to be able to predict the future, so for a while businessmen began to hire economists in large numbers. Eventually, like

the politicians, they were disappointed. Indeed, they became disillusioned with the pretensions of equilibrium economics much more quickly than the politicians. Once they discovered that economists could not predict the fluctuations of the national economy, let alone the output of their own particular industries, businessmen lost interest in economics and in economists.

But as late as the mid-1970s huge fees were paid to forecasting organisations dedicated to supplying businesses with predictions. When Data Resources International (DRI) was purchased by McGraw-Hill in 1979 for US$103 million, it employed more than 500 economists. The demand for forecasts by businesses peaked around the early 1980s, and the companies supplying such forecasts either disappeared or, drastically reduced in size, diversified into management consulting services.

In the early 1980s, Citicorp, then the largest US bank, employed almost 100 people in its economics department. In 1986, Chairman Walter Wriston dissolved the unit. Other major companies including GE, Kodak, Xerox and IBM in the USA, and Midland Bank in the UK, eliminated or drastically reduced their economics departments around the same time. ICI cut their staff of economists from a peak of 30 in 1985 to just five today. Few of the new major corporations of the present day see the need for employing economists at all.

However, in the City of London, on Wall Street and in other financial centres today, the urge to know the future at all costs still overcomes common sense for many people who ought to know better. Where are world stockmarkets heading? At what levels will the FTSE All Share Index, the Dow, and the Nikkei finish at the end of the year 2001? These are the kinds of questions to which fund managers and other investors are desperate to know the answers, and there is no shortage of self-styled experts willing to supply a forecast for a fee. Those City economists famously described by Nigel Lawson as 'teenage scribblers' still sell their forecasts to credulous customers, who would get more accurate answers by making up their own. The one thing which all these forecasters have in common is that, as their track records show (see Chapter 3), they have no idea what the financial markets will do next. Yet, too many people still want to believe that economic forecasting works, and cling to that faith in spite of all the evidence to the contrary. For much the same reasons, snake-oil salesmen were able to ply their trade successfully in the rural United States of the nineteenth century: their customers' needs were more powerful than their scepticism. But most businessmen are not so gullible. Frustrated by the lack of helpful advice they have received from economists, they have turned instead to business schools for ideas as well as for recruits.

Business schools attempt to address specific questions that are of concern to senior managers such as: 'When does it make sense to diversify?' or 'What are the sources of competitive advantage in my industry?', and they address these issues in a language which is readily understandable to their client. The problem of business schools is that their output does not do them any credit in the eyes of their academic colleagues from other disciplines.[1] Most academics share the rather snobbish view that business schools are institutions of intellectual backwardness which provide vocational training for the upwardly mobile. They look askance at the teaching methods, curriculum and research programmes of their business school colleagues.

In many business schools there is a continuing tension among the faculty between the desire for academic respectability on the one hand, and the need to satisfy the corporations who provide their funding, seek their advice and recruit their graduates. Business school students and their potential employers value their MBA degree largely according to the esteem in which the awarding institution is held. An esteemed university will often be reluctant to award a business degree unless the business school faculty behave like their academic colleagues. They want the phenomena of business to be treated in the same way as any other class of phenomena. On the other hand, students who pay substantial fees for their business education don't want to be taught by academic theorists who have never run a business, and resent being asked to study theories that seem no more than intellectual exercises.

Behind this tension lies a fundamental division of opinion about how the subject of business or management should properly be studied and taught. At one end of the spectrum are those who believe that business activity is subject to so many influences and responses that it defies theoretical analysis. Those who believe this developed the case study method of teaching, pioneered by the Harvard Business School. As its name suggests, this is a purely inductive method of learning which appears deliberately to exclude all possibility of theorising. The case study method bears a strong resemblance to the implicit learning approach used by children, in which patterns emerge through trial and error (see Chapter 14).

The widespread adoption by business schools of an explicitly anti-theoretical approach has left the way open for promoters of a succession of 'Big Ideas'. Like meteorites, these flash across the firmament of management thinking for a short time before being forgotten. 'Big Ideas' usually originate in books written by business school professors or management consultants – very often the same people. The books are heavily promoted and sell in large numbers, their surprisingly large circulation often being boosted by bulk purchases by the consultancies which employ their authors. However, they are seldom read beyond the first few pages. These ideas are later disseminated throughout the business community by itinerant lecturers, the best of whom are reputed to command as much as $100 000 for a day's presentation. Such presentations are conducted rather in the style of sermons delivered by evangelical preachers. Ten-point plans to transform your company are mixed with home-spun wisdom and some entertaining anecdotes, the whole mixture being put over with charismatic fervour.

However stimulating such presentations may be, they add to the suspicion that this is not a serious approach to the study of business and that the ideas themselves are no more than passing fads. This is slightly unfair. It is true that the ideas are of highly variable quality. They range from genuinely creative insights offered by thinkers of the calibre of Peter Drucker, Alfred Chandler, W. E. Deming and Charles Handy, to platitudes of stunning banality. But there may be at least two good reasons why such ideas should succeed one another in the forefront of the managerial mind.

First, the business environment is always changing, and it is therefore entirely appropriate that new circumstances should call forth new ideas. So, in the early years of postwar shortages, the task of management was to get the most out of scarce resources;

consequently, ideas of competition and strategy did not figure prominently in the business literature. Later on, in the early 1970s, the idea of competitive advantage captured the imaginations of business leaders. The next Big Idea came along in the 1980s, following the advent of Japanese competition based on flexible manufacturing. This led to the notion of time-based competition, and from there it was a short step to business process reengineering, (BPR). The use of process reengineering to reduce employment drastically led to a reaction, which brought to the fore new ideas about growth and globalisation.

The second reason why business ideas may be so ephemeral is that, when put into practice, they become moves in a continuing competitive game. Once rivals learn to counter them, often by imitation, these ideas will cease to offer a competitive advantage to the firm which initiated them. So businessmen are always on the look-out for the next idea which they hope will give them another, albeit transient, advantage over their competitors.

Even if these two considerations did not exist, it is likely that each idea would eventually disappear because it is invariably partial. It fixes on one small regularity in the stream of business activity, like an eddy in a current, which may have had a certain substance in a particular time and place but which, in the course of presentation, is generalised out of all proportion.[2] One argument commonly used by those who are critical of business schools is that Japan and Germany, for long held out as the great post-war models of business success, flourished without having had any business schools. This argument has lost some of its force as the competitive challenge from Japan and Germany has waned in recent years. Defenders of business schools will also point out that businessmen would not pay out huge sums in consultancy fees to business school professors and in salaries to business school graduates unless they thought they were getting value for money. Many practising businessmen remain sceptical. Speaking about some of his experienced clients, a Taiwanese investment banker told the *Financial Times*,[3] 'For those old guys with hardly any education who built up their companies themselves, theories learned in business school don't mean much.'

A quite different approach to the study of business has been taken by those economists who have specialised in studying management behaviour. They argue that the behaviour of firms and their managers is just as legitimate a subject for academic research as any other type of human behaviour. If the literature emanating from business schools is generally of poor quality then, so the argument runs, that is because the study of business is still in its infancy from the point of view of the application of scientific method. Once a scientific approach is systematically applied to business studies then, it is said, genuine progress will be made in understanding how businesses work. This management-as-a-science approach differs from the traditional business school approach in that it involves looking at business activity as if it were something that takes place at the other end of a microscope. In comparison, the traditional business school approach is to find out what is happening by, as it were, climbing onto the slide, interrogating the specimens, and then trying to help them solve what they perceive to be their problems.

Some management economists believe that by applying equilibrium theory to behaviour within and between firms, they will make discoveries which will ultimately prove beneficial to businessmen themselves. However, despite twenty or thirty years of trying, not a single useful prescription has come out of this activity. But the practitioners of management economics have not given up. One of the most prolific and influential writers belonging to this school of thought is John Kay.[4] His stated view is that the study and practice of management today is in much the same pre-scientific state as medicine found itself two hundred years ago. At that time, doctors peddled universal remedies for all ailments, and prescribed treatments such as bloodletting which very often did more harm than good in complete ignorance of the scientific principles which underlay their subject.

Subsequent progress in medicine came about not only from careful observation of the subjects, but, above all, from the development and application of fundamental knowledge drawn from scientific disciplines related to medicine like physics, chemistry and biology. In the same way, Kay argues, future progress in understanding what makes for good management practice will come about through the application and advancement of our knowledge about economics, sociology and psychology, subjects which he believes play a similar role in relation to management as physics, chemistry and biology in relation to medicine.

Two qualifications must be made to this beguiling analogy. First of all the nature of knowledge in the natural sciences such as physics, chemistry and biology is different from knowledge in the social sciences like economics, sociology and psychology. One major difference is that most of the phenomena being studied in the natural sciences are unchanging, or changing very slowly, while in the social sciences the data is continually evolving in a historical process which is not reversible. Second, physics and chemistry deal with the behaviour of inanimate objects, while the social sciences deal with the behaviour of human beings, behaviour which is made more complicated by the fact of human reasoning and motivation.

These are two of the principal reasons why in the natural sciences there exists a broadly agreed body of knowledge which every competent practitioner must acquire, whereas in the social sciences there does not. This should make us cautious about the prospects for rapid progress in understanding business through the application of knowledge drawn from the social sciences.

The second criticism of the 'management science' prospectus is that what may seem to enthusiasts as being advances in knowledge in the social sciences and in their application to management, may seem to others as representing a step backwards. Take for example some recent developments in equilibrium theorising about business. In the neoclassical equilibrium theory of the firm, each business in the economy is represented as a simple decision-making unit. Given the assumptions of perfect information, the choice to be made is in principle a simple one. It is to choose that combination of technologies and other inputs which will result in maximum profits for the firm. In order to make the optimal choice, there is no need to know anything about such things as the culture of the firm, its internal organisation or even its size. All that is required is to know the prices of all the inputs, and their technical peformance specifications, as well as the price of the output. As all these things are assumed to be known to the firm with certainty, there is no scope for human judgement to enter the firm's decision. The calculation could be done by a computer.

In this theory, only when uncertainty is admitted into the picture is there any scope for human decision-making and human judgement. Of course, once the possibility of human judgement is recognised, there becomes scope for non-optimising behaviour. Survival of the business may well be preferred as an objective to profit maximisation. But once these possibilities of uncertainty and truly human behaviour are admitted, outcomes are unpredictable and the theory breaks down. As it stands, however, it treats the firm as a 'black box', in which nothing is known, or needs to be known about what goes on inside the firm.

This is evidently an unsatisfactory state of affairs, and over the last twenty years various attempts have been made to open the box very slightly without disturbing the conventional equilibrium framework.[5] The possibility of imperfect information is explicitly recognised, but the nature of the missing information is known with certainty as is the cost of acquiring it. Indeed, all transactions have costs which are known with certainty in advance. As Kay admits, these ideas have hitherto been expressed in rather abstruse mathematics, *'even further removed from the day-to-day concerns of practical people running a business'*[6] [emphasis added]. But he believes that what he calls these insights can eventually be translated into everyday business language, and when that happens business studies will have taken a further step in the direction of being a respectable scientific discipline.

This is an illusion. The trouble with these developments is that the comparative static equilibrium method to which they cling is both too limited and too ahistorical to explain more than a tiny proportion of business behaviour. Minimising transactions costs in a given context entirely ignores the processes of growth within the firm, specifically the way in which new choice sets evolve. It also fails to comprehend the interaction over time between organisation and technology, a relationship which is critical to understanding the long run development of most firms.[7] As Casson has observed, while transactions cost economic theory has succeeded in explaining where the boundaries of the firm might be drawn, it has nothing to say about the much more interesting question of how these boundaries relate to what goes on inside the firm.[8]

If two rival companies are operating within exactly the same business environment, why does one company do better than another? For students of business, this is one of the most fundamental questions of all, but the equilibrium model and its derivatives cannot address it. Neoclassical theory assumes that all firms are in fact the same whereas all the evidence is that different firms behave differently even when operating in the same environment.[9] The reasons why similarly circumstanced companies behave differently include each company's distinctive capabilities (what it can do that others can't), how far these capabilities fit into the ever-changing needs of the market, and the extent to which they can be imitated by competitors. Another reason why outwardly similar companies differ in their behaviour is that each one 'sees' their market differently. In other words, while each may act 'rationally' given their perceptions, their behaviour and priorities are likely to differ markedly as a result of differences between them in their perceptions of their business environment. This is an example of the influence of the distinctively human factor in economic activity.

The image of the firm as a desiccated optimising calculating machine portrayed by neoclassical equilibrium theory has coloured economists' perception of business behaviour. For example, there is a tendency to believe that major strategic decisions must be the product of exhaustive calculation. More often, they may be the result of panic, greed, opportunism or even the regulatory environment, which itself may be in the process of changing in response to the last 'botched' takeover.

Likewise, there is a tendency to accept that excellence in business performance is something for which there must exist an ascertainable formula of universal applicability. Once implemented, the formula can be maintained indefinitely. In fact, the observed fluctuations in the performances of large companies may more plausibly be accounted for by the proposition that the ever-changing environment at some periods favours their peculiar intrinsic strengths, while at other periods it requires aptitudes or competences which they lack. Only very rarely do large companies possess as a 'core competence' the ability to adapt quickly to almost any environment.

Mechanical models, by their nature, cannot take account of the fact that each company has a unique identity, with a life and culture of its own. The sheer diversity of company behaviour eludes the crude classification offered by these models. Once we recognise that each company may be a unique entity, then we must consider whether biological rather than mechanical analogies might not offer a better understanding of economic activity. Biological analogies have a long history in economics and business. Alfred Marshall, the great English economist whose working life spanned the last two decades of the nineteenth century and the first two decades of the twentieth century, was fond of drawing an analogy between the position of firms in an industry and the position of trees in a forest. Business theorists also frequently have recourse to biological metaphors to get their points across. For example, Gary Hamel, a professor at London Business School, recommends increasing the 'genetic variety' of a firm by recruiting senior managers from a range of different backgrounds.

Quite recently, some new ideas have been put forward which further develop the notion of firms and industries as living systems. In the next chapter, we shall see how the perspective of an economic organisation as a living system represents an improvement upon the perspective which sees it as a mechanical system.

6
From Mechanical to Biological Analogies

> 'The economy responds to changes in policy not so much like a machine responding to a control, but more like an animal being poked with a stick.'
>
> Paul Ormerod, *The Death of Economics*

Economists as well as politicians, businessmen and financial journalists are accustomed to using mechanical metaphors whenever they want to describe the effects of monetary or fiscal policy in a contemporary advanced economy. Terms such as 'overheating' and 'slowdown' are commonly employed, and it has recently become fashionable to speak of the need for a new 'financial architecture' for the world economy. Within large nation states, particularly in Western Europe, it is taken for granted that governments are largely in control of economic events, and that this control is achieved by a 'touch on the throttle' or an 'application of the brakes' to achieve a 'soft landing'. The only judgement required of the controller is about the extent and timing of these adjustments. But the use of these mechanical metaphors suggests a degree of government control over the behaviour of a market economy which, while conveying an effortlessness and finality congenial to many minds, is wholly inappropriate.

In the passage quoted above, Paul Ormerod observes that an economic system responds to changes in government policy not so much like a machine responding to pressure on a throttle or a pull on a lever but more like an animal being poked by a stick. It is impossible to know exactly how it will respond. Although familiarity with the behaviour of a particular animal might allow the handler to learn what to expect in certain circumstances, in the market economy, unlike in stock-handling, circumstances do not repeat themselves.

When one thinks of a firm as a living organism or a whole economy as a living system, the implications are very different from thinking of them as machines. In the mechanical perspective, the firm's task is seen as *optimising* something (profits, shareholder value) by exercising management *control* from the top down. In the biological perspective, the objective of the firm is *survival* and that is achieved through successful *adaptation* to its environment. In economic theory, biological analogies have a long history. One of the earliest and most celebrated is Mandeville's *Fable of the Bees* (1714), in which the behaviour of bees was used to illustrate what the author described as the paradox of private vices and public virtue. He was drawing attention to the observation that purely selfish behaviour on the part of individuals can sometimes, quite unintentionally, give rise to benefits for society as a whole. This is a proposition we shall return to in Chapter 7. Quite recently, Ormerod has drawn an analogy between the behaviour of ants in a colony and individuals in human society.[1] Both are shown to be examples of complex systems, in which unpredictable behaviour in the short run gives rise over time to a form of regularity.

In the eighteenth century, a French physician turned economist, François Quesnay, whom some regard as the founding father of modern economics, drew up his famous 'Tableau Economique', depicting the flows of expenditure between the different sectors of an economy. His work was quite openly modelled on the concept of the flow of blood through the human body.

At the beginning of the twentieth century, in the Introduction to his *Principles of Economics*, Alfred Marshall declared forthrightly that 'The Mecca of the economist lies in economic biology rather than economic dynamics'.[2] He then went on to explain that because 'biological conceptions are more complex than those of mechanics' a book such as his which was intended to give an introduction to the subject of Economics would have to 'give a relatively large place to mechanical analogies'.[3]

Why are biological conceptions to be preferred to mechanical ones in trying to analyse economic activity? There are several reasons. First, the process of motion in a mechanical system is reversible, and does not give rise to qualitative change. In a biological system, the processes of motion are irreversible, and represent a cumulative sequence of adjustments that can and often do give rise to qualitative changes. Mechanical systems treat human beings as if they were inanimate particles condemned to respond passively to events in obedience to immutable laws. The environment in which they operate is assumed to be unchanging, and characterised by perfect information and global rationality. In fact, human behaviour in economic as in other areas of activity, is usually purposeful and frequently creative. Each individual strives to achieve something in an environment of which they have very incomplete knowledge, and where rational behaviour (that is the consistent and calculated pursuit of well-defined objectives) is therefore unattainable. There is the possibility, even the likelihood of persistent error-making, and thus of non-optimising behaviour. Finally, chance plays no part in mechanical systems, but it has a very important role in the evolution of biological systems, as it does in human behaviour.

Biological systems are much richer and more complex than mechanical ones, exhibiting, in Hodgson's words, 'continuous change, tangled structures and causalities, and huge variety'.[4] In short, any satisfactory explanation of economic phenomena must address interactions between human beings, who are living creatures, and cannot limit itself to stochastic outcomes and mechanical cause and effect. Biological analogies, while an improvement on mechanical ones, also have their limitations. There is no counterpart in economics to the Mendelian theory of sexual reproduction, nor is there anything corresponding to the differentiation which exists between different species. Conversely, there is nothing in the biological world corresponding to the inheritance of wealth amongst human beings.

None of these limitations would appear to constitute an overwhelming objection to the adoption of biological analogies. What probably makes most economists reluctant to give up the mechanical metaphor is that it means giving up the capacity for precise prediction.[5] However, as the world of human economic activity has proved to be unpredictable in any specific or detailed sense (see Chapter 3), this must be counted as a gain, not as a loss. As we shall discover in Chapter 7, a limited degree of what may be called 'pattern predictability' in complex non-mechanical systems may still be possible.

Of course, not all biological analogies for economic behaviour are necessarily helpful: some may be misleading. For example, Alchian, Friedman, Williamson and others have tried to invoke the concept of natural selection in support of the mechanical hypothesis of profit maximisation. Their proposition is that, if the process of market competition is perceived as a 'Darwinian' struggle between firms, in the long run only efficient behaviours such as attempted profit maximisation will survive. There are two implicit assumptions in this proposition: one is that profit maximisation in the neo-classical sense is what actually happens in the real world. The other is that profit maximisation represents efficient behaviour. This proposition comes perilously close to what may be called a Panglossian view of the world, in which the observed existence of any pattern of behaviour is taken as evidence of its efficiency. In fact, many biologists recognise that biological evolution is often influenced by historical accidents which can lead it to take a sub-optimal path through time. Likewise, many economic historians accept that the course of development of an economic system at any time is in large part influenced by the path it has followed in its recent past. Surviving institutional arrangements are not necessarily the most efficient just because they exist. We shall return to this question of path dependency in the next chapter.

Sometimes a biological analogy can be so elaborate and amusing that we become seduced by it, and therefore tend to overlook its limitations. Such is the case with the fable of 'The Tortoise and the Hare', recounted by John Kay.[6] His story concerns a group of tortoises who once upon a time lived happily in some marshes, a hostile environment for their natural predators. But the tortoises grew envious of their neighbours the hares, who lived equally happily on a nearby plain, where their speed afforded them some protection from their predators. The fable concerns the desire of the tortoises to match the speed of the hares, an ambition which eventually had an unhappy ending. The moral which the author wants his readers to draw from this fable is that those firms which are successful in business are those whose characteristics fit the environment in which they operate. The tale is told in such an entertaining fashion that we tend to forget an important limitation on the usefulness of the analogy, namely that while animals can alter neither their own characteristics nor their environment, firms, being collections of human beings, can do both.

This brings us to the decisive objection to the use of biological analogies in understanding the behaviour of economies and of firms; that is, that there are fundamental differences between human beings and other living organisms. While human beings do indeed conform to all the rules of animal behaviour, they are capable of additional functions, many of which are critically important for the purposes of understanding the nature of economic activity. These functions include the capability to make plans, which may include the formation of quite complex expectations about the responses of other human beings, and to be able to exercise judgement in executing these plans.

Whereas cultural evolution in animal communities is almost insignificant,[7] in human societies it is of central importance. There is no counterpart in the animal world to the transmission of wealth by inheritance or to the transmission of knowledge through organised instruction. Unlike biological evolution, which is Darwinian, human cultural evolution is Lamarckian. So in human societies acquired characteristics can be transmitted from one generation to the next. And human cultural evolution proceeds very much faster than biological evolution.

This is not to deny that there exists learning and decision-making processes in animal communities. Indeed, there are branches of biology devoted to the study of such processes from which lessons may be drawn which can be helpful in understanding some corresponding aspects of human behaviour. But it is important to emphasise that when we draw on such studies we are not dealing with analogies but with the same processes, albeit on a less complex scale.[8]

Recently, there have been some serious attempts to draw lessons for learning behaviour in human organisations like firms
by studying adaptive behaviour in groups of animals and birds.
Amongst the most notable of these contributions is that of Arie
De Geus.[9] His argument runs as follows:

The key to survival in biological systems is adaptation to a changing
environment. In human social organisations like firms, as in biological systems, successful adapters learn to develop new skills
to exploit the environment in new ways. Any firm which aims
to survive for a long period of time will experience fundamental
environmental change over which it has only limited control.
Like a living organism it has no choice but to change as best it
can to restore its balance with its environment.

De Geus refers to studies of bird behaviour by the late American
biologist Allan Wilson, who concluded that species of birds that
flocked together, like bluetits, learned new behaviour faster than
others which did not, like robins. While robins are territorial
birds, bluetits are gregarious. Robins are antagonistic in their
behaviour to one another across the boundaries of their territo-
ries, and therefore have fewer opportunities to learn from a change
in each other's behaviour. Bluetits, on the other hand, learn more
quickly from each other to exploit new opportunities in their
environment. A particular example of their more rapid learning
behaviour was the introduction of metallic milk bottle tops in
suburban England in the 1920s. Bluetits quickly learned from
each other how to penetrate the new tops: the robins took much
longer.

As with birds, so with human beings, writes De Geus. Even if
a firm contains a number of innovative individuals, any new
ideas they may introduce are less likely to catch on unless there
is a process for transmitting skills from the individual to the
group, and the relevant individuals must have the opportunity
to move round the group. Thus the rate at which a business
organisation learns will depend on its practices in areas like team
working, job mobility and career development. De Geus attaches
great importance to organisational learning as a factor explain-

ing a firm's performance. He remarks that a firm's ability to learn faster than its competitors may be its only sustainable competitive advantage.

Instructive though such comparisons may be, and it is important to repeat that we are talking not about analogies but about the same processes, we must bear in mind their limitations. Business firms can take initiatives to alter the environment in which they do business through such measures as price-cutting, political lobbying, mergers and acquisitions. The capacity of non-human living creatures to alter their environment is much more limited. Birds and fish are condemned to respond to their environment in a wholly reactive manner. They do not try to change their environment, nor do they choose a particular long-term objective. It is true that some species of birds and fish are capable of performing remarkable feats of migration, but neither their destination nor their route are the result of making conscious choices. Businessmen on the other hand do make conscious decisions all the time. Some of these decisions may be based in large part on 'hunch' or 'intuition', which may be a sub-conscious form of pattern recognition, but others involve a calculated choice between objectives or destinations for their businesses.

Furthermore, human beings that are in control of businesses, whether large or small, usually want their business to do more than just survive. They have personal goals or they set corporate objectives which they would like their business to achieve. It does not much matter whether these objectives and the strategies for reaching them exist only in the minds of individual entrepreneurs or whether they are the product of a corporate planning team. It does not matter whether these consciously planned strategies succeed or fail, or even if they are wholly misconceived. The important point is that the behaviour of the firms concerned is strongly influenced by the formation and attempted implementation of such plans, that is by the purely human element in their composition. And, as knowledge grows and spreads, it seems likely that the specifically *human* ingredient in economic and business behaviour will grow more important as time goes by, at the expense of other influences.

To accept uncritically the comparison of business firms with other living systems would be to limit a company to behave in a purely reactive manner in response to threats and opportunities occurring in its business environment. It would be to ignore the extent to which the distinctly human element in business behaviour is the decisive additional factor.

Of course, it may be argued as De Geus in fact does, that for a company to formulate a strategic objective and to pursue it through thick and thin may be prejudicial to that company's long-term survival. It may be that in the very long run, adaptation to its changing business environment is the only realistic option for a company if it wishes to survive, and that the pursuit of any other objective is no more than vanity and self-deception. That would seem to be the conclusion which De Geus draws from his study of long-lived companies. However, there is at least one human organisation, namely the Papacy, which has pursued single-mindedly an overriding objective, namely the propagation of the Faith, while at the same time succeeding in surviving for several centuries.

Because the behaviour of human societies more closely resembles that of animal communities than it does the behaviour of a machine, a well-chosen biological analogy is likely to be more illuminating in a didactic sense than a mechanical one. But the limitations of biological analogies must also be recognised.

Human beings differ from animals in their greater ability to reason, to act purposefully, to learn, and to communicate with each other. These differences lead to another important characteristic which distinguishes human beings from other creatures namely the greater range of their ability to alter their environment or to become independent of it. Furthermore, the human capacity to alter our environment is increasing. Our ability to transform life, not just by ancient and slow procedures like animal breeding but by modern methods like genetic engineering, will soon be so great that the future of life on earth will come to depend in large part on crucial choices made by the human species. As Murray Gell-Mann has observed, it looks as if the role of natural biological evolution in the foreseeable future will be secondary, for better or for worse, to the role of human culture and its evolution.[10]

As an improvement on analogies, scientific studies of the behaviour of groups of animals and birds can sometimes provide lessons for understanding the corresponding processes, such as learning behaviour, in human organisations. But the corresponding human process is always likely to be more complex simply because it involves interactions among human beings. In the next chapter, we shall see that all observed phenomena can be arranged in a hierarchy of complexity, and that a class of processes known as complex adaptive systems can be used as a framework to provide a better understanding of the behaviour of human and non-human organisations alike.

7
The Economy as a Human Complex Adaptive System

'Wherever we look, we discover evolutionary processes leading to diversification and increasing complexity.'
G. Nicolis and I. Prigogine, *Self-Organisation in Non-Equilibrium Systems*

In 1964 the Austrian economist Freidrich von Hayek suggested that one could rank the different branches of science in a hierarchical order according to the degree of complexity of the phenomena they sought to investigate.

A hierarchy of complexity

Hayek observed that the phenomena encountered in interactions between human beings were more complex than those found in the interactions between physical bodies. Because of this, human behaviour is both more difficult to explain and more difficult to predict than the behaviour of physical bodies. This perhaps accounts for the famous remark of the German physicist Max Planck who in the 1930s told the economist John Maynard Keynes that as a young man he had thought of studying economics but he had eventually abandoned the idea as he realised that it would be too difficult.

So, in studying human behaviour (that is the province of the social sciences) we should not expect to achieve that predictability of detail and uniqueness of outcome which has been achieved in the physical sciences. This is not because the social sciences are more backward, but because the phenomena being studied in the physical world are much simpler. Indeed, the fundamental laws of physics are very simple indeed. When Hayek sketched out his hierarchy of the different branches of science, physics, being the simplest, was at the bottom, with biology in the middle and the social sciences dealing with the most complex phenomena, at the top.

Quite recently, the idea of a hierarchy of phenomena of increasing complexity has been taken up by some natural scientists, including the Nobel prize-winning physicist Murray Gell-Mann (1994). Gell-Mann extended Hayek's classification to include chemistry so that the hierarchy now reads, in ascending order of complexity – physical, chemical, biological and social phenomena. Gell-Mann also recognises the existence of artificial (that is computer-generated) phenomena, although it is not clear where he thinks they belong in the hierarchy. The principle which governs the ordering of classes of phenomena in the hierarchy is that, at any given level, a class of phenomena will obey all the laws governing their own behaviour *as well as* all the laws governing the behaviour of classes at lower levels in the hierarchy. Those higher up exhibit an additional complexity of behaviour which justifies distinctive methods of analysis and the formulation of separate laws of behaviour. Therefore, the behaviour of all living organisms conforms to the laws of biology, as well as obeying the laws of physics and of chemistry.

Consider what happens when a stone is thrown into the air. A stone obeys only the laws of physics, and its behaviour is therefore quite predictable. If a bird is thrown into the air, it is subject to exactly the same physical forces as the stone, but additional forces – chemical and biological – are at work, which account for its more complex behaviour. If a man-made vehicle is launched into space it is capable of even more complex behaviour than a bird. Being controlled by a human mind, it can execute manoeuvres to attain precisely planned objectives.

Complex phenomena and prediction

The importance of distinguishing between those classes of phenomena which are simple and those classes which are more complex can be seen very clearly in the question of prediction. In the case of physical phenomena the corresponding laws can very often be stated as functions of a few variables whose values can be ascertained. Prediction is therefore relatively straightforward. Because of the remarkable progress which has taken place in the physical sciences and in its related technologies within the last century, people have supposed that similar progress should be possible in the biological and social sciences. The problem with these more complex classes of phenomena is that no person can acquire all the facts which would be necessary, according to the corresponding theories, to produce predictions of particular events.

Hayek illustrates this proposition with reference to the Darwinian theory of the evolution of biological organisms.[1] It would be possible to obtain a complete explanation of the structure of all existing organisms if it were possible to find out all the particular facts of the past which influenced the selection of the specific forms that emerged. Likewise, it would be possible to predict the future development of an organism, if it were possible to know all the particular influences which would operate on that organism in the future. Neither of these sets of facts can be known, so the prediction of specific events in biological phenomena is not possible. Nevertheless, even although the *specific* results of the mechanism of selective evolution will depend on the unknown contextual conditions in which it operates, predictions of a more general character, which might be called pattern predictions, can be made. Furthermore, these pattern predictions are capable of falsification.

What is true of biological phenomena is true, *a fortiori*, of the still more complex phenomena of human social interaction. While the prediction of particular events in human social systems such as market economies is not possible, it may be possible to undertake successfully the prediction of outcomes in the form of patterns.

We said in the last chapter that it is the distinctively human element in human organisations like business corporations and whole economies which gives their behaviour a greater degree of complexity than is found in corresponding biological organisations. But what exactly do we mean by complexity?

What is complexity?

Although more sophisticated definitions are on offer,[2] the degree of complexity of any system can be loosely identified with the number of relationships of interdependence between agents within a given system. For example, the number of non-zero cells in an input–output table could be taken as an approximate indicator of the complexity of the economic system to which it refers. Scholars seem to agree that the degree of complexity of the human social as well as of the physical and biological world is continuously increasing.[3] Kauffman makes the same point more colloquially when he remarks that there are increasingly numerous ways to get lunch.[4]

Gell-Mann points out that as time goes on, higher and higher forms of social complexity keep appearing.[5] This observation is quite consistent with a decrease in complexity in any given social system. For example, the consequences of unusually severe climate changes or outbreaks of external or internal strife might force a retreat to simpler patterns of social interaction. In the extreme case, a society might disappear altogether.

Complex systems

Complex systems are systems defined by the interaction of large numbers of agents. Not only human, but also biological, chemical and even physical systems may be defined as complex in this sense. The common factor in complex systems is that the interactions between the agents are *non-linear*.

It might be supposed that non-linear interactions between large numbers of agents, to which is added the periodic intervention of chance, would produce outcomes characterised by confusion and disorder. In fact, as we shall see in the following chapters, many complex systems exhibit regularities which make them worthy of study. Such systems were at first given the name 'adaptive non-linear networks' by John Holland,[6] but they have since come to be better known as 'complex adaptive systems'. It is usual to give as varied examples of such systems the behaviour of investors in financial markets, the operation of the immune system in mammals, the behaviour of organisms in ecological systems, and cognitive and learning processes in animals and in humans.

In fact, complex adaptive systems are quite pervasive. As Hayek observed, *all* enduring structures above the level of the simplest atoms and up to the brain and society are the results of, and can only be explained in terms of, processes of selective evolution leading to increasing diversity and complexity.[7] The persistence of complex systems which have emerged from evolutionary processes is to be explained by the continuous adaptation of their internal states to changes in the environment. The individual elements or agents within these systems possess sufficient regularity of conduct, or the capacity to follow rules, so that the results of their actions tends to restore the pattern of the system if it is disturbed by external influences. In other words, the persistence of complex adaptive systems is to be explained by an interaction of patterns representing the principles of self-organisation and evolution, and not by a mechanistic conception of uni-directional laws of cause and effect.[8]

Non-linearity

The implications of non-linearity are profound. It can easily be shown that two otherwise identical non-linear systems that have very small differences in their initial conditions will quite quickly follow very different paths through time. This means that although in principle the outcomes of non-linear systems are determinate, in practice they are not. And if we introduce chance into the systems, in the shape of periodic disturbances, then the outcomes will be quite unpredictable in detail.

The weather provides a familiar example of a physical complex system. Very slight differences in environmental conditions can cause any given weather system to develop in quite different ways. Weather systems are therefore unstable, and so their behaviour is unpredictable in detail. From the point of view of the user who is looking for precision, weather forecasts are notoriously unreliable. But the weather is not wholly unpredictable. Regular patterns in the behaviour of the weather are quite familiar.

While no one can predict with confidence how much snow will fall in central London on Christmas Day 2001, we can be confident that snow in central London is more likely on Christmas Day than on Midsummer's Day. As we shall see, regularity of pattern but not of detail is a characteristic feature of complex systems.

The importance of chance

The importance of chance in biological systems is well-understood and represented by genetic mutation, but the importance of chance in the economic system has not really been recognised by equilibrium theory. One interesting exception concerns the origins of the Great Depression of the 1930s. One of the most famous propositions of Milton Friedman and Anna Schwartz in their authoritative *A Monetary History of the United States*[9] is that the failure of the Federal Reserve Bank to prevent a shrinkage in the quantity of money in the US banking crisis of 1930/31 turned what might have been a minor recession into a major depression. Friedman and Schwartz attribute the failure of Fed leadership to the premature death in 1928 of Benjamin Strong, the energetic chairman of the New York Fed. As Brittan observes, it is remarkable that Friedman, who has in his other work put so much emphasis on the primacy of impersonal market forces (that is on equilibrium theory), should attribute critical importance to one historical accident.[10]

Anyone who has been in business for a long time and who reflects on his experiences can hardly fail to be impressed by the important part which pure chance has played in both his successes and his failures. Missing a flight can make the difference between winning and losing a contract. If the contract is important, the whole future development of the firm, and of the industry of which it is part, will be different from what it would have been but for that chance event. The reader can reflect on the sequence of chance events which led to his parents meeting each other. Without these, you, the reader, would not be reading these words.

The importance of chance in human history has of course long been recognised. It is memorably in the cry of Richard III wanting to escape from the lost battle of Bosworth Field, 'My Kingdom for a horse'. The lesson of this episode has been spelled out for generations of schoolchildren: 'For want of a nail, the shoe was lost, for want of a shoe, the horse was lost, for want of a horse, the crown was lost'.

Chance plays an important part in the evolution of biological systems, too, and even in the much slower evolution of physical systems, so that history can be seen as largely the result of a sequence of what Gell-Mann calls 'frozen accidents'.[11] Were it not for chance, the history of the world would be quite different from what it is, and the future would be different from what it might otherwise be.

All this should be quite familiar to us, and yet, strangely enough, there is no place at all for chance in traditional economic theory which treats economic activity as if it were no more than a simple mechanical process. Even in the recently fashionable chaos theory, it seems that chance need play no part. Most definitions of chaotic systems indicate that the apparently random behaviour exhibited by such systems is a consequence of their non-linearity.[12] The outcomes of non-linear systems are in principle determinate, even although in practice they are not because of their sensitivity to small differences in initial conditions.

The outcomes of interactions within complex systems, on the other hand, cannot be determined exactly in advance. This is not only because of their non-linearity, but also because in addition they are subject to the influence of so many chance factors. From this perspective, much of the effective complexity which we can observe in the world today is attributable to chance. An important difference between linear and non-linear systems is that in linear systems chance disturbances are always dampened. Following a disturbance, the system will always return to equilibrium. This is sometimes known as a negative feedback loop. A simple example is that in the event of a sudden increase in the demand for a commodity its price will rise, thereby encouraging an increase in the supply of the commodity to restore the balance between supply and demand.

Non-linear systems, on the other hand, have positive feedback loops as well as negative ones. Positive feedback loops occur when the effect of an initial chance event is amplified rather than dampened. An example is a business start-up where a modicum of initial success creates a degree of market credibility enabling the business to achieve still further success.[13] In economic theory, the existence of potential economies of scale provides an opportunity for positive feedback. In such circumstances, expanding output leads to lower unit costs, which if reflected in lower prices brings an opportunity for further expanding output as a result of increased market demand.

The combination of positive and negative feedback loops, allied to the interjection of chance events, means that complex systems exhibit periods of order and stability interrupted by unexpected periods of self-generated instability. This is precisely the pattern of episodic behaviour which Douglass North says has characterised the long run development of market economies.[14]

Non-linear systems also differ from the more familiar linear systems in the relationship between the parts and the whole. In linear systems, the behaviour of the whole system is a simple sum of the behaviour of the parts. This makes analysis quite straightforward, and no doubt accounts in large part for the widespread use of mechanical methods. In complex systems, however, where the component parts undergo simultaneous non-linear interactions with one another, the behaviour of the whole system is no longer the sum of the behaviour of the parts. Furthermore, the aggregate behaviour of the system often feeds back to the individual parts, modifying the behaviour of some of them. For example, reports of a sharp upsurge in house prices may arouse widespread expectations of a forthcoming rise in interest rates, which may in turn cause some individual businesses to adjust their behaviour.

Adaptive behaviour

Adaptive behaviour can be defined as any process whereby an entity is progressively modified to give better performance in its environment.[15] In the way that they change over time, both biological and human systems are adaptive. Each agent is constantly acting and reacting to what the other agents in the system are doing as well as reacting to changes in the common external environment. An important distinction between the classes of phenomena at the bottom of the hierarchy of complexity, that is the behaviour of physical and chemical systems, and those higher up the hierarchy, that is biological and human systems, is that the latter are capable of adaptive behaviour, while the former are not.

Adaptive behaviour in biological systems is guided in part by acquired experience. Every living creature has encoded in its genes an implicit prediction: 'in situation ABC, action XYZ is likely to pay off'.[16] Cats, for example, have inherited certain patterns of behaviour which help them to catch mice. The more sophisticated is the organism, the greater is the part played in its behaviour by individual learning compared to the 'theories' programmed into it by biological evolution.

Physical systems, on the other hand, are not adaptive. While in biological organisms, the experience represented by past variation and natural selection are handed down to future generations of the organism, no such factors are at work in physical systems. Although large eddies in a weather system or in the current of a turbulent river may appear to 'give birth' to smaller ones downstream, no information is transmitted or processed in such systems, therefore they are not truly adaptive.

In human complex adaptive systems such as the market economy, participants learn from experience how to play the game more effectively, whether the game is competing with other companies in the supply of financial services to households, or whether the game is simply shopping at the local supermarket. In most markets, suppliers will have to anticipate how their competitors might respond if they undertake certain actions. As participants adapt in an attempt to survive in an environment in part provided by the behaviour of other participants, the pattern of interactions will evolve over time. In other words, the rules of the game will change as a result of the behaviour they produce. As John Holland has observed: 'It's essentially meaningless to talk about a complex adaptive system being in equilibrium: the system can never get there. It is always unfolding, always in transition.'[17]

Adaptive behaviour in individual human beings is sometimes called 'learning', and this phrase has also been applied to adaptive behaviour in human organisations and societies. Individual human beings can learn in two ways: through formal instruction and through 'learning by doing'. For participants in a market economy, learning by doing is by far the more important type of learning. Just how important it is will be seen in the next chapter where we discuss entrepreneurship.

Why is learning by doing, in other words learning through trial and error, so important in human economic behaviour? It is because, contrary to one of the fundamental assumptions of equilibrium theory, whether they act as producers, traders or consumers, people do not find themselves in possession of all the relevant information they would like to have when it comes to making a decision. They are usually very far away from being in that happy situation, so optimisation in practice is out of the question. And complexity and chance make discovery of the optimum an inappropriate as well an impossible task. The best that can be hoped for is successful adaptation to continuously changing circumstances. Better performance in this respect means exploiting information as it is acquired, and the only way of acquiring information in most circumstances is by testing the available options. Trial and error is the defining characteristic of adaptation in all complex systems, whether human or non-human. In the human complex system which is the market economy, competition is a process of discovery in which new knowledge is acquired and transmitted.[18]

In the biological world, increasing complexity and diversification is the result of the random mutation of genes in the phenotype. In the world of human behaviour, it is largely the outcome of the peculiarly creative capabilities of human beings. In the particular area of economic activity, human creativity principally takes the form of entrepreneurial behaviour. The unceasing attempts by entrepreneurs to make a profit by introducing and exploiting new ways of doing things result in a market economy which is characterised by perpetual novelty. Growth in the contemporary market economy is not typically a question of supplying more of the same types of commodities using the same processes of production, as equilibrium models suggest. Rather,

it is a question of new and different commodities being produced, distributed and consumed, using new and different methods.
Perpetual novelty is not just a peripheral feature of the advanced
market economy: it is one of its defining characteristics.

Perpetual novelty

Perpetual novelty is also a characteristic feature of complex adaptive
systems in general. The very fact of adaptation implies qualitative
change: that is, novelty. In biological systems the rate of adaptation
is relatively slow: in contemporary human systems it is relatively
fast. Complex adaptive systems typically have many niches, each
one of which can be exploited by an agent adapted to fill it.
The very act of filling one niche opens up others, so new opportunities are always being created by the system. In biological
systems, opportunities arise for new parasites, new predators, and
new symbiotic patterns. In human systems, opportunities continuously arise for new consumer and capital goods, new methods
of production and distribution and new forms of organisation.

Chance and path-dependence

Complex adaptive systems are said to exhibit path-dependence.
This means no more than that the future development of a system
is influenced by the path which it has followed in the recent
past.[19] Applied to a biological system, this is unremarkable. Applied
to a human economic system, the proposition ought also to be
unremarkable, but it needs to be emphasised since it contradicts
another assumption of equilibrium theory. Equilibrium theory
implicitly assumes that, whatever its starting point, a system will
always find its way to the same (optimal) point of equilibrium,
and so history can be ignored. But economic historians have
pointed to cases in which demonstrably sub-optimal solutions
have been adopted simply as a result of the prior occurrence of
arbitrary choices or chance events. The examples given include
the selection of standard formats in railway gauges, computer
technologies, video systems, and the 'QWERTY' format of the
typewriter keyboard.

 These examples of 'technology lock-in' have been questioned

both on grounds of their historical accuracy and their empirical significance. If that were all there was to path-dependence, the concept would be worth no more than a footnote in any text on economic theory. But for economic historians like North, path-dependence is much more than the proposition that yesterday's choices are the starting point for today's. For him, the basis of path-dependence is Hayek's notion of collective learning, which is the key to understanding the historical success or failure of human societies to adapt their economies in response to changing circumstances.[20]

According to North, the cultural heritage of any society includes the institutions, that is formal and informal rules, which have been formed on the basis of past experiences. These rules are used to filter and interpret current experience. In cases of successful economic development, the information from current experience is interpreted in ways which lead to modification of the rules so that current problems are resolved and/or economic performance is improved. North emphasises that there is no simple set of rules that will lead to 'successful' outcomes in all circumstances. On the contrary, rules that will be successful in one context may be obsolete in another. In any given situation, more than one set of rules may provide the appropriate incentives for productive activity, and rules themselves must adapt to new circumstances.[21] So the success of the rules is context-dependent. This is a perspective quite different from the universalist and ahistorical one afforded by equilibrium theory.

The Hayek–North proposition about path-dependence in economies remains a conjecture, albeit an important one. However, some statistical evidence is available to support the hypothesis of path-dependence at the level of the firm.[22] In a recent survey of the available econometric evidence on the growth of firms, Gerowski concludes that corporate growth is a path-dependent process. In other words, increases in firm size are driven by unexpected (unexpected by the econometrician) firm-specific shocks which have permanent effects on the size of the firm.

Two other implications of the data are worth noting. Corporate growth rates appear to be idiosyncratic. The evidence that there is a tendency to a common growth rate across firms in the same industry is 'pretty weak'.[23] In cyclical downturns, it appears that many firms remain substantially unaffected, while some actually prosper. Gerowski draws the conclusion that 'corporate growth is history dependent, and every firm seems to have its own particular history.'[24] It is very difficult to reconcile the evidence on which this conclusion is based with any of the equilibrium theories of the firm (see Chapter 5).

Optimisation

In complex adaptive systems, improvement is more important than optimisation. Agents can never optimise: the space of possibilities is too large, and there is no practical way of finding the optimum. As we noted above, a complex adaptive system is always unfolding. When parts of the system do settle down to a local optimum, it is usually temporary. Should they remain at that equilibrium for an extended period, then those parts are described by Holland as almost always 'dead' or uninteresting.[25] Even in the rare cases when a relevant global optimum can be defined, the system is typically so 'far away' from that optimum that basins of attraction, fixed points, and the other apparatus used by equilibrium theory in studying optima can tell us little or nothing about the system's likely behaviour.

Self-organisation

Since complex adaptive systems represent the interactions of a myriad of agents amongst each other and with their common external environment, it might be supposed that the result would be confusion and anarchy, a criticism which Marx and his followers levelled at the market economy. But this is not the case. What makes complex systems so interesting is not only that they normally exhibit some regular patterns of behaviour, that is some degree of order, but that this order emerges spontaneously without any form of external or internal control. This phenomenon of the spontaneous emergence of order out of apparent confusion is known as *self-organisation*.

Self-organisation can formally be defined as the process whereby an initially unorganised system acquires increasing self-control in a complex environment as a result of the spontaneous interaction of agents, without there being control by any one agent or without any external control on the system.[26] In a huge variety of circumstances in the physical, chemical, biological and human social worlds, complex patterns of behaviour emerge from systems characterised by very simple rules. The biggest example is the universe itself, 'the full complexity of which emerges from simple rules plus the operation of chance'.[27]

In the world of biology, the principle of self-organisation is illustrated by the behaviour of birds as they move through the air in a flock, avoiding obstacles and other threats and seeking insects or other opportunities. Shoaling behaviour in fish demonstrates the same principle. The orderly behaviour of such groups is noteworthy in two important respects. First, coordination is achieved without any control being exercised either from inside or outside the group. Experiments with artificial (computer-simulated) groups confirm that even the most complex patterns of group behaviour can be achieved if each individual member of the group follows a few simple guiding principles. Second, the coordinated and orderly behaviour of the group as a whole is beneficial to the individual members.

Likewise, in the trillions of transactions between millions of individuals and firms which make up the typical modern market economy, an overall outcome emerges which is both orderly and beneficial at the level of the economy as a whole as well as at the level of the individual market. This outcome was neither intended nor planned by the participating individuals, who are merely following their own particular interests. The emergence of this order or pattern represented by the network of markets which we call the market economy is another example of self-organisation. Like the emergence in history of such other human institutions as the law, language, money, and even the state itself, it may be said to be the result of human action, but not of human design.[28]

Evolution

Complex adaptive systems evolve. When they interact with an environment which is also evolving, they are said to co-evolve. The difference between the evolution of complex systems and the dynamic change of physical or mechanical systems can be understood in the following way. A dynamic system is a fixed pattern in time, expressed in the form of difference or differential equations. Evolution means a change in the pattern, and in the parameters of the equations describing the system. In complex adaptive systems, the parameters can be thought of as changing continuously and irreversibly.

The market economy as a complex adaptive system

If we were to attempt to summarise those features of a market economy which characterise it as a complex adaptive system, they would be as follows:[29]

Outcomes are the result of the interactions of large numbers of individual agents who are not necessarily homogeneous. The behaviour of any given agent depends in part on the expectations that that agent forms about the behaviour of a number of other agents. These expectations which agents form about the behaviour of other agents are unlikely to be explicit, coherent or mutually consistent. Whereas in equilibrium theory all agents are fully-informed rational (that is consistent) optimisers, in the complexity perspective agents have to make sense of their problem, learn and adapt. They do not optimise, not because they are constrained by a finite processing capability,[30] but because the concept of an optimal course of action is meaningless in a world complicated by the actions of other agents, and one which is ever changing. In general, different agents know different things, their knowledge being acquired from experience mediated by specific cognitive processes.

There is no external controller in the market economy. Individual behaviour is constrained by formal and informal rules, as well as by the consequences of the process of market competition. The complex system that is the whole economy is composed of innumerable smaller complex systems operating at different levels. At any given level, organisations, behaviours, strategies and commodities all serve as 'building blocks' for the corresponding units at a higher level.

A modern market economy exhibits all of the principal characteristic features of a complex adaptive system, namely complexity itself, evolution, perpetual novelty, path-dependency, self-organisation and the influence of chance. This suggests that using the framework of a human complex adaptive system may help us to understand better how a market economy really works. But complexity theory provides only a framework for thought. The task of the remaining chapters is to put some theoretical flesh on these bones. The theories to be deployed are not new. We shall see how some important ideas of both Classical and Austrian economists are congruent with the paradigm of complexity.

8
The Coordination of Economic Activity

'By directing that industry in such a manner as its produce may be of the greatest value, he intends only his own gain, and he is in this, as in many other cases, led by an invisible hand to promote an end which was no part of his intention. Nor is it always the worse for the society that it was no part of it. By pursuing his own interest, he frequently promotes that of the society more effectually than when he really intends to promote it.'

Adam Smith, *The Wealth of Nations*

Adam Smith seems to have been the first economist to recognise that the market economy was a self-organised system, although he did not of course use that phrase, and the first to appreciate the benefits of self-organisation. He observed that: 'It is not from the benevolence of the butcher, the brewer or the baker that we expect our dinner, but from their regard to their own interests'.[1] And every individual who invests his own capital where it will achieve the highest return is thereby contributing 'to render the annual revenue of the society as great as he can', even although 'he intends only his own gain'.[2] In modern language, what Smith can be interpreted as saying is that the independent actions of individuals in a society inadvertently give rise to desirable overall results in the shape of an intelligible social process which is beneficial in the sense of increasing the level of income of society as a whole. This proposition can be divided into three parts.[3]

The first is that human action often has unforeseen and unintended consequences. The second part is that the sum of these unintended results aggregated over a large number of individuals and over a period of time results in an orderly process which looks as if it had been consciously designed even although it has not. The third part of the proposition is that the resulting order is beneficial to the participating individuals in ways they had neither anticipated nor intended. In the case of a network of markets this benefit arises principally from the coordinating functions which the markets perform.

Despite the absence of any central control or direction, a market economy presents a relatively orderly pattern of activity. It is necessary to say 'relatively' because of the more or less continuous disturbances represented by corporate and personal bankruptcies, factory closures, losses of employment, and the volatility of prices in financial and primary commodity markets. There are also the periodic fluctuations in aggregate economic activity known as business cycles, which we discuss in Chapter 11. But the outcome of the interactions of millions of people each pursuing their own individual interests is not anarchy but order. What exactly is the nature of this order in a market economy?

In general, competition between the participants in a market economy brings about a state of affairs in which (a) everything will be produced which someone knows how to produce and which can be sold profitably at a price at which buyers will prefer it to the available alternatives; (b) everything produced is being produced at least as cheaply as could be achieved by those not producing it; and (c) everything will be sold at prices lower than those at which they could be sold by anyone else not doing so.[4] These may seem like modest accomplishments, and there is certainly nothing about them which could be described as 'optimal', except perhaps there is no known way of bringing about better results. It should also be noticed that the order or pattern thus defined is benignly responsive to the continuing changes in consumer tastes, technologies and the availability of resources which characterise all contemporary market economies.

How is this result brought about? As Smith pithily observed, wants are not satisfied directly. The butcher does not provide the wherewithal for Smith's evening meal because Smith wants it, but as a consequence of the fact that he (the butcher) is trying to make a profit. Likewise, today's manufacturer of breakfast cereal or provider of internet services does not supply a commodity to satisfy the needs of a known individual, but because he hopes that thousands of individuals whom he does not know will buy the commodities in question.

Adam Smith posed, but could not answer, the question which many believe remains the central question for economic theory. How does it come about that the behaviour of large numbers of individuals, each pursuing their own interests and equipped with limited knowledge, often erroneous and largely tacit, results spontaneously in the emergence of a coordinated outcome, and that the resulting order is robust and adaptable in the presence of continuing changes in resources, tastes and technologies?

The pattern of order which is observed in market economies results from the acquisition and utilisation of knowledge by the participants. In the process of competition, new knowledge is uncovered, and that knowledge is then communicated from one participant to another principally, but not exclusively, by changes in prices. The essence of a market economy is that it is a way, better than any known alternative, of utilising the knowledge which is dispersed amongst participants, knowledge which is usually fragmentary, incomplete and even contradictory. The assumption of equilibrium theory, that there is a given unchanging stock of knowledge accessible to all participants or available in one place, simply assumes away the fundamental problem of organisation in any economic system, which is how to achieve a coordination of wants and supplies.

How does market competition contribute to the utilisation of knowledge? One way of understanding competition is to think of it as a game, the outcome of which for each player will be determined by a mixture of skill and chance. As in other competitive games, information is discovered in the process. In sporting events and in examinations, it is discovered who is best on a particular occasion. In the market, it is discovered which goods are wanted and which are not, and who is the lowest cost supplier of these goods.

Competition can also be compared to experimentation in science.

It is primarily a discovery procedure. And it is not just a question of utilising knowledge which is dispersed amongst hundreds of thousands of people. It is a question of giving them the incentive to use their abilities to discover such facts as will be relevant to their task in a particular situation. This is not of course the mechanism of competition to be found in the textbooks of equilibrium theory, in which producers can sell as much as they like without being able to influence the market price, and where the quality of goods as well as the technology with which they are produced is the same for all producers. The Austrian concept of competition, entirely consistent with the complexity perspective, is that it is a process in which producers compete with each other through the periodic introduction of qualitatively improved products using improved methods of production and distribution, as much as they compete through price.

How competition leads to coordination

People are unaware of each others' plans, including their expectations concerning prices.[5] If buyers and sellers are optimistic, that is if sellers set too high a price for buyers, both will be disappointed and some will have to revise the plans if there is to be any trade at all. If they are too pessimistic then sellers will begin by setting prices that are too low in relation to what buyers would be prepared to pay. In that case different prices may appear for the same goods, and the prices of some goods may exceed the costs of the bundles of resources that are required to make this. In both cases, there arise opportunities for buying cheap and selling dear which attract the attention of entrepreneurs. By exploiting these opportunities for profit, the effect of entrepreneurial activity is to bring prices of goods into line with the costs of their constituent resources.

As part of this same process, resources will be shifted from less-valuable to more-valuable uses (measured by the prices consumers are willing to pay), more-productive technologies will replace less-productive ones, and new sources of raw materials will be discovered.

If there were not continuing changes in consumer wants,

technologies and available resources, then entrepreneurial activity would soon bring about a situation where throughout the market for any good only a single price would prevail. The prices of goods would be equal to the prices of their constituent resources and all potential buyers would be able to buy what they want at the price they expect to pay. But of course, in the real world of incessant change these corrective tendencies will be interrupted or even frustrated for a while. Despite that, the coordinating tendency of entrepreneurial activity is unmistakable.

This implies a view of competition which is very different from that held by neoclassical equilibrium theory. According to the latter, competition is *a state* of equilibrium, and a prerequisite of that state is the existence of perfect knowledge throughout the market. Without perfect knowledge, competition is said to be imperfect, and the whole analytical apparatus of equilibrium theory collapses like a house of cards.[6] But for classical economists – that is for both Austrians and complexity theorists – competition is *a process* through which knowledge is discovered and communicated. When an entrepreneur enters a market with a better product, a new process of production or simply a better value proposition, he is inspired by the hope of a profit. But only when he takes action does he discover the true situation, that is what the reaction of buyers really will be, what costs really are. When he makes this discovery, the result will be passed on to the rest of the market, not only through the changes in prices which his actions have set in motion, but by his example being followed by competitors. In other words, competition is a process of trial and error, in which the principal agents, the entrepreneurs, proceed on the basis of learning by doing. The contrast between these two views of competition could hardly be greater[7]

Knowledge is communicated by the price system. Even when some relevant facts become known only to a few, the effects of their actions on prices will influence the decisions of others. Thus speculators perform the socially useful function of ensuring that relevant information is passed on by an appropriate change in price. Competition not only gives anyone who has the opportunity to exploit special circumstances the possibility to do so profitably, but by conveying in coded form the information that some such opportunity exists, the competitive game secures the utilisation of widely dispersed knowledge.[8] Suppliers will substitute one input for another in their processes of production or distribution on the basis of the knowledge of one simple fact, namely that the price of one input has fallen relative to the other. It is not necessary for them to know all the changes throughout the rest of the economy which on balance have resulted in this price change. Nor does the consumer need to know all the materials and their costs which have gone into the supply of his breakfast cereal. All he needs to know is the price of the commodity itself and its suitability to his needs.[9]

The competitive process continuously adapts the use of resources to conditions unknown to and unforeseen by most people. By not preventing unforeseen changes from spreading their effects, but by facilitating adaptation to such changes, the market helps to diminish avoidable uncertainty. Of course, the adaptation to the new situation will never be perfect.[10] Furthermore, given that circumstances unforeseeably change, some people's expectations are going to be disappointed. Such an outcome will be as much the result of chance as of skill. In Hayek's words: 'It is one of the chief tasks of competition to show which plans are false'.[11]

Prices are not the only way by which information is communicated in markets. Suppliers learn from direct observation which of their competitors' new products appeal to buyers.[12] Buyers, in turn, learn from the experience of frequent purchases which suppliers are trustworthy: that is, those which can be relied upon not to exploit information asymmetries that may from time to time arise. Competition is not only a method of utilising the knowledge and skills of others, it also encourages each individual to acquire knowledge and skills of their own. It encourages the development of a commercial spirit in a society, and 'rational' behaviour and entrepreneurship are as much the products as the conditions of effective competition. Hayek puts it quite bluntly: 'Competition is . . . a process in which a small number makes it necessary for larger numbers to do what they do not like, be it to work harder, to change habits.'[13]

Hayek also observes that: '. . . by guiding the actions of individuals by rules rather than specific commands it is possible to make use of knowledge which nobody possesses as a whole'.[14]

What are these rules, and where do they come from? This is the topic we address in the next chapter, on the Evolution of Institutions.

Before leaving the coordinating function of markets, however, it is perhaps worth pointing out that the principle of self-organisation offers a more satisfactory understanding of Adam Smith's famous metaphor of the Invisible Hand than does the concept of a general economic equilibrium, proof of the existence of which is often claimed as a vindication of Smith's metaphor. Equilibrium theory assumes implicitly that, with perfect information, all markets can clear instantaneously and simultaneously. In other words, the coordination problem is simply assumed to have been solved. Furthermore, the conditions for achieving equilibrium impose severe constraints on the behaviour of individual agents. As Vriend points out, each individual is assumed to take the structure of the equilibrium model into account in calculating his choices, thereby *anticipating* the equilibrium character of the overall outcome.[15]

By contrast, the complexity approach seeks to show how the apparently spontaneous formation of regularities in the economy as a whole can be *explained* by the actions of individual agents following simple rules of behaviour. The challenge is to explain the emergence of such regularities as the unplanned coordination of trade in decentralised market economies. The fact that the agents' interactions are non-linear means that resort must be had to simulation methods if quantitative results are required. But this approach which seeks to explain how aggregate level regularities arise 'from the bottom up' through the repeated local interactions of autonomous agents is surely preferable to the imposition of fictitious 'top-down' coordination devices such as market-clearing constraints.[16]

9
The Evolution of Economic Institutions

'The essential point to grasp is that in dealing with capitalism we are dealing with an evolutionary process.'
J.A. Schumpeter, *Capitalism, Socialism and Democracy*

When Schumpeter wrote these words he did not mean 'evolutionary' in the Darwinian sense of a biological process of natural selection. He was referring to an older and broader meaning of the word 'evolution' which refers to development from within. In his own words, the development of an economy involves 'changes in economic life as are not forced upon it from without but arise by its own initiative from within'.[1] Schumpeter wanted to emphasise that the behaviour of a market economy through time was to be understood in terms of its own internal developments, and not as the result of a sequence of external shocks, the approach taken by equilibrium theory. In other words, we should look for an explanation of events like the East Asian financial crises of 1997 or the stockmarket crashes of 1987 in New York and in London in terms of prior developments within the economic systems concerned.

All complex adaptive systems are evolutionary in the sense that they seek to explain the movement of a system through time in terms of how it arrived at its existing position from a previous stage. So, in the language of complex systems, economic activity in a market economy is carried on within a social and natural environment which changes, and, by its changes, alters the data of economic activity. Both the agent and the environment are at any point the outcome of the last process. In the case of human systems, the actions of agents significantly alter the environment in which they operate.

Complex adaptive systems including the market economy exhibit regularities of pattern and diversity of detail. While we can hope to identify some patterns of behaviour which are common to most if not all market economies, we cannot expect to find any commonality of detail. This should not be surprising. Human interactions are so complex, and the influence of chance so pervasive, that the chain of cause and effect between events through time is usually long, uncertain and impenetrable. It is difficult, if not impossible, to distinguish its elements even with hindsight.

For example, it seems to be generally accepted by historians that the Wall Street Crash of 1929, and the subsequent Great Depression were at least in part a result of the economic upheavals brought about by the First World War. Yet few scholars would care to delineate precisely the exact sequence of the intervening chain of events. Likewise, it is possible to sketch an argument linking the collapse of the fixed exchange rate between the Thai baht and the US dollar with the fall of the Berlin Wall some eight years earlier, but it is impossible to be conclusive. Again, it was possible to predict correctly that the entry of a new competitor, Direct Line, bringing a new method of distribution and a new technolgy to the motor insurance market in the UK in 1985 would lead eventually to mergers among some of the existing suppliers. But it would have been impossible to have predicted which companies would have merged or when such mergers would have occurred. In fact, Royal Insurance merged with Sun Alliance, and Commercial Union merged with General Accident, both in 1998.

As we saw in the last chapter, complex adaptive systems manifest continuous adjustments to changes in their environment, demonstrating the principle of self-organisation. They are also the result of processes of selective evolution. Complex systems are always evolving. What exactly does 'evolution' mean in the context of the complex adaptive system which is the market economy?

An evolutionary process can be defined as one in which the

present state of the elements can be explained, at least in part, by their previous state. Or, as Boulding puts it: 'Everything is what it is because it got that way: any object at any moment of time is what it is because of its previous history and its laws of growth'.[2] The essential idea of evolution therefore has two charac- teristics: change that comes from an earlier state in historical time, and change that comes from within. The contrast with movement in a dynamic equilibrium system is clear. The move- ment of a ball across a billiard table does not depend on the earlier state of the ball. Any change in its motion can only come from forces outside the ball. Likewise, the response of an equi- librium model of an economic system to a change in one of its variables is quite independent of the previous history of that system (indeed, it is fully reversible). And the change can only come from outside the system: it must be 'exogenous'.

To most people, the concept of evolution is inseparable from biology. This association has perhaps been one of the main ob- stacles to the acceptance of its applicability to the phenomena of human society.[3] (Although there seems to have been no com- parable resistance to the application of concepts drawn from classical mechanics.) In fact, the origins of the concept of evolu- tion can be traced at least to the social philosophers of the eighteenth century. Darwin is believed to have got the idea for his famous theory from reading Malthus.[4]

In a curious parallel development in the twentieth century, the concept of the complex adaptive system nurtured by Hol- land and other natural scientists at the Santa Fe Institute in New Mexico in the 1980s seems to have been anticipated by Hayek about 30 years earlier.[5] Yet it is still widely believed that com- plexity theory is something which has developed from the natural sciences, and should therefore be treated with caution in appli- cations to the social sciences!

The theory of biological evolution has undergone some formal- isation since Darwin, and contains some well-defined elements such as individuals, species, populations, genotypes and pheno- types. It also retains the fundamental Darwinian proposition that evolutionary change only occurs directly through changes in the genome, although indirectly through the differential fate of pheno- types feeding back upon the distribution of genotypes. In simple terms this means that no matter how many generations of mice have their tails cut off, mice will continue to be born with tails.

Economists and others who have wished to apply the principle of evolution to the social sciences have recognised that the mechanism of transmission is quite different. In human societies, acquired characteristics, most importantly knowledge and skills, *can* be transmitted from one generation to the next. So human cultural evolution proceeds much faster than biological evolution. In human beings, innate knowledge, that is, that part of knowledge which is genetically transmitted, is very small in relation to that which is learned, and therefore in human societies the influence of cultural evolution dominates the influence of biological evolution. Amongst non-human animals, the position is just the reverse. Although there are occasional examples of learned knowledge causing changes in behaviour in animals and birds (recall the example of bluetits and robins in Chapter 6), these are generally swamped by innate influences.

These fundamental differences between human and non-human animals justify assigning them to different levels in a hierarchy of complex phenomena. Yet some economists still persist in searching for counterparts in the domain of human society for all those elements which appear in the Darwinian theory of biological evolution. Not finding them, they conclude that evolutionary theory is inapplicable to the social sciences.[6] A similar type of error is perpetrated by 'social Darwinists', who focus on the selection of individuals rather than institutions, and on the innate rather than the culturally transmitted capacities of those individuals.

A theory of cultural evolution

The process of human cultural evolution shares three characteristics with the process of biological evolution: a mechanism of transmission, a mechanism for selection, and the influence of chance.

Human societies have developed over the generations from primitive bands of hunter-gatherers numbering perhaps up to 40 people, led by a headman and defending a territory against all outsiders, to civilised societies of today, numbering tens or

hundreds of millions. Whatever their size and level of development, the behaviour of individual members of each society has
been and still is governed by formal rules and informal norms
of behaviour, customs and conventions, which collectively can
be defined as institutions. According to Hayek, these institutions
can be divided into three classes, according to their origin.[7] There
are first of all those institutions that arise from innate feelings,
genetically transmitted. Then, second, there are those institutions that are the result of deliberate calculation or rational choice.

But the most important institutions of modern society like
language, law, money and markets, and even government itself
fall into a third category. All of these institutions have evolved
as a result of a long process of collective learning, whereby practices which were introduced into a society perhaps by chance
but which turned out to give that society a differential advantage were adopted and passed on in the culture. There is in this
case a process of selection of institutions according to their success in furthering the interests of the group, with subsequent
transmission by human communication. In truly primitive stages
of society, communication could take the form of imitation, as
in learning processes in animals. In modern times, the introduction of arrangements for formal education has enormously
accelerated this process of transmission.

As societies progressed, said Hayek, new rules had to be learned
to make possible the coordination of the activities of larger, more
successful, groups. At the earlier stages, advances were achieved
by adopting rules which infringed or repressed natural biological instincts. Later, however, changes in the rules took the form
increasingly of relaxation in prohibitions, with the result of increasing the freedom of the individual. Such practices as the
toleration of barter trade with outsiders, the recognition of limited private property, the enforcement of contractual obligations,
competition with fellow craftsmen in the same trade, the lending of money with interest, were all infringements initially of
older rules. In Hayek's words:

> ...the law-breakers who were to be path-breakers, certainly
> did not introduce the new rules because they recognised that
> they were beneficial to the community, but they simply started

some practices advantageous to themselves which then did prove beneficial to the group in which they prevailed.[8]

Hayek is anxious to emphasise that the principal institutions of modern society, including the market economy, are the outcome of this long process of cultural evolution, and should not be seen as the product of some acts of deliberate collective decision-making, or the application of reason. He quotes Adam Ferguson:

> The forms of society are derived from an obscure and distant origin; they arise, long before the date of philosophy, from the instincts, not from the speculations of man . . . We ascribe to a previous design, what came to be known only by experience, what no human wisdom could foresee, and what, without the concurring humour and disposition of his age, no authority could enable an individual to execute.[9]

and '. . . nations stumble upon establishments, which are indeed the result of human action but not the result of human design'.[10]

Hayek goes on to argue that the rules which underpinned the market economy were learned by a population consisting chiefly of independent farmers, artisans, merchants and their servants and apprentices who shared the daily experiences of their masters:

> They held an ethos that esteemed the prudent man, the good husbandman and provider who looked after the future of his family and his business by building up capital, guided less by the desire to be able to consume much than by the wish to be regarded as successful by his fellows who pursued similar aims.[11]

Then, Hayek appears to echo Schumpeter's analysis of the decline of pure capitalism when he writes that 'an ever increasing part of the population grow up as members of large organisations and thus as strangers to those rules of the market which have made the great open society possible'.[12] As a result, says Hayek, long-submerged animal instincts have come to the top. We pursue these issues in Chapter 15. Meanwhile, two comments are in order.

First of all, while it may be conceded that up to about a century ago societies were 'better served by custom than by understanding'[13] (that is by learning to do 'the right thing' without knowing why it was the right thing), it seems clear that from quite early in the twentieth century the application of new knowledge in the form of industrial technology began to have an increasing influence on the development of the advanced market economies, and indeed on the behaviour of advanced societies generally. In the twenty-first century, this 'directed' component of evolution is likely to have an even more powerful influence. It is true of course that the consequences of this development remain in part unforeseeable, but it does seem that in future those institutions which have been consciously designed may play a more important role in the evolution of the market economy.

Second, Hayek chides traditionalists and conservatives for believing that the norms of behaviour of a society are, or at least ought to be, immutable. He believes they are always evolving: 'Tradition is not something constant, but the product of a process of selection guided not by reason but by success'.[14] Despite his admiration for the virtues of the market economy and its associated political order of representative government, it therefore seems as if Hayek does not share the point of view of Fukuyama, for whom the market economy is an institution from which there is no foreseeable evolution.[15]

The economic historian Douglass North has worked out a fully-articulated model (he modestly does not claim that it is a theory) of social and economic change, in which the influence of Hayek is clearly discernible. The details of his model vary slightly from one presentation to another, but the essential features may be summarised as follows:[16]

The socioeconomic process may be regarded as a competitive game, in which the players are organisations (governments, business firms) and individuals. The objective of the game is survival, and the rules are what North calls institutions. These include both formal laws and regulations and informal norms of behaviour, each of which have their corresponding sanctions. Competition between organisations produces events. But events cannot be observed objectively; they are perceived and interpreted according to the beliefs held by the players. For any player at any moment, their set of beliefs comes from three possible sources: (1) inherited beliefs, (2) what they have learned from experience, and (3) formal instruction. The second and third sources arise because organisations find it advantageous to acquire knowledge to gain competitive advantage over their rivals.

As a result of beliefs altering, institutions also change, and these institutional changes in turn have an influence on events. North illustrates his model with some historical examples. In the seventeenth century, the power of Spain declined. Contemporary documents show that this decline was well-understood, and numerous reforms of economic policy were proposed. None, however, was adopted, showing that there is no guarantee that economic and political decline will induce perceptions on the part of the influential players that will lead to the reversal of that decline. It remains to be seen whether the evolving perceptions of voters and politicians in Russia today will produce the changes in norms of behaviour which will bring about a desired improvement in economic performance in that country.

North remarks that those countries in Africa and Latin America which have failed to develop have failed because their inherited belief systems – their cultural heritage – discouraged the evolution of modern economic and political institutions. Markets remain primarily local. As they become exposed to outside forces, the existing institutions result in organisations whose profitable opportunities are to be found in re-distributive rather than in productive activities. In these cases, governments become little more than 'theft machines'.

Perhaps the experience of the UK in the last twenty-five years provides yet another illustration of North's model, this time with a happier outcome. In the late 1970s, postwar economic performance was perceived to be poor relative to competitor countries. After a number of remedies had failed, the belief system accepted the proposition that redistributive coalitions were too powerful, damaging overall performance. Once the powers of the trade unions had been reduced, and much of the public sector privatised, productivity and economic performance improved. Thus British economic and political institutions in this period of history proved to be adaptable.

The experience of the Japanese economy in the last nine years suggests that that country has been less fortunate. The institutional arrangements which served it well in the immediate postwar period, a high degree of central coordination and regulation, albeit informal and consensual, seem no longer to be appropriate to present circumstances. The necessary institutional changes have been proceeding very slowly, perhaps because the affluence of the society means that the need for change seems less urgent. Likewise, contemporary France and Germany have been slow to reform many of their economic institutions, notably those relating to their labour markets and social security systems, even though neighbouring European countries have exemplified the nature of the required reforms.

The co-evolution of institutions

Hayek thought in terms of the institutional selection process operating at the level of the group rather than the individual. Within a group, rules of behaviour for the individuals emerged not because of the advantages to the individual but because they improved the chances of survival of the group as a whole. Subsequently, other writers have argued that selection takes place at the level of the individual. Each individual, for example, may be thought to select from among a set of rules which compete with each other.[17] But these arguments are not mutually exclusive: evolution can take place simultaneously at both levels. This is an example of what is called co-evolution. It is a phenomenon which can also be seen in the evolution of markets.

It is apparent that institutions like money have evolved because of the evident convenience to all members of a society in having a common medium of exchange instead of having to barter with each other. Likewise markets came into existence, it is conjectured, because, amongst other advantages, they reduced the costs of search and negotiation for both buyers and sellers. The existence of rules, however informal, gives both buyers and sellers confidence in participation in a market, a vital although intangible and therefore often overlooked ingredient in executing a successful transaction.

Loasby points out how the evolution of markets can guide the evolution of goods and services traded in those markets.[18] The rules of any market (and rules can differ between markets) provide a framework within which variation amongst the goods supplied by different sellers can be tested by the responses of consumers. Such responses can take the form of direct comments as well as the information conveyed by their buying decisions. They provide the basis for what Loasby calls 'informed conjecture – though never rational choice'[19] about new products and services. At the same time, markets allow consumers the opportunity of discovering new ways of meeting their needs, and even of recognising new needs. This in turn provides opportunities for innovations by suppliers. So Loasby concludes that:

> ... a market is not an arena for the co-ordination of pre-defined supply and demand functions, but an institutional setting for the cognitive processes by which supply and demand are continually reshaped.[20]

Furthermore, some product innovations may require the creation of new market institutions, so that economic evolution shows evolution within market institutions as well as the evolution of these institutions themselves.

Economic institutions like markets also co-evolve with political and other institutions in society. In all advanced economies, governments impose a regulatory framework within which markets are obliged to operate. These politically imposed regulations are more restrictive in some societies than others and, within societies, some markets than others. At the present time, however, there does seem to be a general trend towards more restrictive regulation of markets throughout the Western world. However, an examination of any particular market will normally reveal a co-evolution over time between the regulations and events in the market concerned.

For example, the market for retail financial services in the United Kingdom was relatively lightly regulated until the Financial Services Act of 1986 came in to force. The purpose of this Act was ostensibly to 'protect' consumers of financial services; it attempted to do so by laying down detailed rules for the processes by which sales should be conducted. Even before the Act was replaced thirteen years later by the Financial Services and Markets Act of 1999, the regulatory authority had radically altered its initial rulebook as a result of the experience of its operation.

Meanwhile, over the same period the processes of the market itself had evolved as a result of new entrants, new products, and new technologies, each of which underwent their own processes of evolution. The new Act of 1999 attempted to respond to these changes as the authorities perceived them. As a result, the method of regulation has shifted in emphasis from regulation of the sales process to regulation of product quality, and the stated objectives of regulation have altered. A process of continuing mutual adaptation between regulations and events in the market has thus occurred and can be expected to continue. This process is typical of such interactions between political and economic institutions throughout the market economies. Nor should it be imagined that this process will eventually converge to some equilibrium. On the contrary, the advent of the Internet and other electronic means of distribution is likely to have major disruptive consequences for regulatory practices in this and many other markets.

Of course the evolution of new norms of behaviour is not confined to market behaviour. It is a perfectly general phenomenon throughout the whole area of human behaviour, as Fukuyama illustrates in describing the consequences of the introduction of a new traffic law in Washington DC.[21] This suggests the possibilities for complex adaptive systems providing a common methodological approach for a unified social science.

Evolution and complexity

As we saw in the last chapter, opportunities for increased complexity keep arising in all complex systems. The result is a tendency for the maximum level of complexity in any system to drift upwards. In biological evolution, complexity can either increase or decrease at each evolutionary step, but the effect on the whole set of existing species is that the greatest complexity represented has a tendency to grow larger with time. In economic evolution, the tendency for complexity to drift upwards is illustrated by the pattern observed when a society grows richer. Any individual family may from time to time see its income rise or fall, while the range of incomes is growing wider, so that the size of the largest family income tends to keep increasing.[22]

In the evolution of market economies, when opportunities for making a profit arise as the result of the coming into existence of a new product or new technology, it is likely (but not certain) that an entrepreneur will eventually come along and realise them. This step corresponds to filling a niche in biological evolution. In both cases, opportunities may remain unfilled for some time, although not forever. Once the opportunity is realised and the niche is closed, the path of evolution thereafter is different forever from what it might otherwise have been. The environment is irreversibly altered for all actual and potential participants. It is easier to observe these unfilled niches in economics than it is in biology. Boulding gives the example of transportable batteries for storing electricity, for which the payoff has remained large yet the niche unfilled for almost a century.[23]

Increases in complexity come about through cooperation as well as competition. In the biological world, a transition from a lower to a higher level of organisation can come about through the formation of a composite structure, as in the evolution of multi-celled plants and animals from single-celled organisms.[24] Likewise, in social systems, individuals can get together to form business organisations, and business organisations can cooperate with each other through trade associations and through links with governments.

Is the tendency to increasing complexity inevitable?

Gell-Mann says that in non-adaptive (physio-chemical) complex systems like galaxies, the movement to higher maximum complexity proceeds in a manner that resembles diffusion, the outcome of an essentially random process. In complex *adaptive* systems, however, it seems that selection pressures often favour higher complexity, although there is no reason to believe that this is a general rule. Indeed, he describes the task of characterising those systems and environments where complexity is highly advantageous as an important intellectual challenge.[25]

Boulding identifies increased complexity with adaptability, and makes an important distinction between adaptability and adaptation.[26] In the biological world, sudden major environmental changes, such as those which separated one geological age from another, eliminated many species which had been *well-adapted* to their previous environment. Those which survived were, on the other hand, the more *adaptable*, even though they may have been less well-suited to the earlier environment. Such 'gateway events' as major environmental changes may open up huge new opportunities for adaptable organisms, leading them to acquire 'new and very significant regularities' and raising them to a higher level of complexity.[27]

Another possible explanation for the observed pattern of increasing complexity runs as follows: an ecosystem may have an unfilled niche for a species just above the level of those which constitute it. While mutations that do not increase complexity mean that the resultant organisms have to compete with the well-established occupants of existing niches, those mutations that represent a move towards increased complexity, and especially those towards adaptability, may find new niches in existing systems that seem rather full.[28]

Punctuated equilibrium

A well-known phenomenon of advanced market economies documented by economic historians is that their growth over the long term has been neither continuously smooth nor continuously volatile. Instead, long periods of relative stability are interrupted at irregular intervals by periodic upheavals.

A similar phenomenon appears to characterise biological evolution, although the process is so much slower that it is difficult to observe directly, and resort must be made to fossil records. In the biological world, it seems that species remain relatively unchanged for long periods of time, and then undergo comparatively rapid change over a brief period. Two explanations have been put forward. One is sudden dramatic changes in the physical environment. The other is that a small change in the genome, coming on top of a long series of earlier changes, can trigger one of the major changes that punctuate the relative stability of biological evolution.

Conclusions

Complex adaptive systems are always seeking new opportunities, filling niches, creating new ones at the same time, and experimenting with novelty. They proceed by trial and error. They occasionally stumble on 'gateway events' that open up the possibility of new regularities, including new complex adaptive systems. In biological systems, these gateway events usually result in significant increases in complexity.

Human beings are not only the most complex living organisms, but we have now developed the power to control other living organisms through genetic modification. This means we can alter the course of biological evolution on this planet. Whether we make the right conscious choices, and whether our culture will evolve the necessary institutions, that is the necessary rules and norms of behaviour, to allow the planet to remain both habitable and civilised remains to be seen.

10
The Lessons of History

'The rationality assumption of neoclassical economics
assumes that the players know what is in their self-interest
and act accordingly. Ten millennia of human economic
history say that this is a wildly erroneous assumption.'
Douglass North, 'Some Fundamental Puzzles in
Economic History'

What lessons can be learned from history about the nature and
progress of economic activity in human societies? The highlights
are simply stated. Human beings separated from other primates
around four million years ago. Agriculture began around ten
thousand years ago, while economic growth in the modern sense
began some four hundred years ago. It is this last period which
has been intensively studied by economic historians who wish
to learn something about the process of growth in market economies. Most are agreed that around two hundred years ago a
sequence of events which can be called the Industrial Revolution
began in parts of Western Europe. As a result, industry displaced
agriculture as the major activity in the advanced economies.
Somewhere around fifty years ago, also in the advanced economies,
industry began to yield in importance to services in a development
which some have labelled the Information Revolution.

Looking at the whole of recorded human history, North observes that economic progress has been episodic. Periods of
sustained growth have been exceptional: periods of stagnation

or decline have been more common. Until 1950, economic growth in the modern sense of the development of a market economy with private ownership of capital was confined to a small part of the planet. He concludes that:

> Economic history is an endless depressing tale of miscalculation, leading to famine, starvation, deceit and warfare, death, economic stagnation and decline, and indeed the disappearance of whole civilisations. Even the most casual inspections in today's news suggest that this is not purely an historical phenomenon.[1]

If we look at the origins of economic growth in the early modern period of Europe some four hundred years ago, we find a number of states competing with each other politically while at the same time interacting with each other through trade and the exchange of ideas and practices. In this decentralised, competitive situation each society faced a slightly different external environment, and each pursued different policies. Some societies (England, the Netherlands) succeeded in sustaining their development, while others (Spain, Portugal) failed, but even the experience of the failures contributed to the successes of the others. North stresses that it was the pursuit of a variety of options (as opposed to a single Europe-wide policy) in the context of strong competition which produced a creative environment out of which some policies and practices evolved which were conducive to economic and political development. Although the Netherlands and England pursued different paths to success, a common pattern can be identified in both cases. Social beliefs were shaped by external events so as to induce the evolution of political and economic institutions favourable to economic growth. In both cases, the competition they faced from other evolving nation-states was both an important underlying source of change and a constraint on the options open to their rulers.

The story of the evolution of market institutions has been well-established by historical research.[2] The codes of conduct which underpinned long-distance trade in the Middle Ages eventually formed the basis of commercial law; the development of marine insurance (and life assurance) converted some kinds of uncertainty into risk; the development of the bill of exchange and of early banking companies provided the basis for a capital market. Such innovations in institutions and organisations created in the Netherlands, by the end of the sixteenth century, the first modern economy. In other European countries it was also true that the rising costs of warfare compelled monarchs to cede power to early forms of representative assemblies in exchange for tax revenues. But in England and in the Netherlands, circumstances led to the further development of representative government, whereas in much of the rest of Europe there was a 'persistence or revival of absolutist regimes with centralised decision-making over the economy'.[3]

This divergent pattern of institutional development was transferred to the American continent. The English colonies in North America continued such informal norms of behaviour as the traditions of religious diversity and local control which they had inherited from the parent country. Amongst the formal legal rules, 'fee simple' ownership rights in land, and secure property rights in other markets were carried over. On the other hand, the Spanish and Portuguese colonies in South America had thrust upon them a uniform religion and centralised bureaucratic control. According to North, this created incentives for organisations and entrepreneurs (both economic and political) to pursue redistributive rather than productive activities. The subsequent very different histories of North and South America are a vivid illustration of the importance of path-dependence.

In general, concludes North, 'the gradual development of informal norms of behaviour that have become deeply embedded in the society provides the stable underpinning' to the adaptability of institutions which is a feature of those Western economies that have experienced sustained growth in the modern period.[4]

This conclusion is somewhat disappointing if one's objective is to discover a 'formula' for promoting sustained economic growth which could be transferred to those parts of the world which have not experienced it hitherto. One is reminded of the story of an American visitor to an English stately home who enquired of its owner if he could have the recipe for producing such a fine lawn. 'It's quite simple', came the reply, 'just lay down some grass seed and roll it for five hundred years.'

Is there any way that formal institutions can be substituted for informal ones that are absent or inadequate? Again, the experience of Latin America is not very encouraging. When they became independent from Spain in the early nineteenth century, many of her former colonies adopted a political constitution modelled on that of the United States. Subsequent developments both politically and in terms of economic growth have led to very different outcomes from those in North America.

The recent experience of the newly industrialising countries (NICs) of East and Southeast Asia and of countries like China and Russia suggests that there is no general institutional 'formula' for sustained economic growth: informal norms as well as formal rules may be context-specific. If that is so, it might explain the apparent inability of Western advisers hitherto to create the requisite conditions for growth outside the areas in which it has already occurred.

Not that the advanced economies can afford to be complacent. The late Mancur Olson drew attention to the danger of what he called 'redistributive coalitions', for example trade unions, strangling the continuing growth of productivity in mature economies.[5] When this happened, or threatened to happen, as it appeared to do in the UK in the 1960s and the 1970s, corrective action followed. Will contemporary Germany and France take similar steps? As North puts it, the key question for any society is this:

When the growth of productivity in the economy is adversely affected by internal redistributive coalitions or by changes in external competitive conditions or by changes in technology is the political organisational structure sufficiently flexible to change the framework of institutions so as to restore the competitive position of the economy?

A noticeable feature of postwar economic growth in the advanced economies has been the enormous and continuing expansion of formal rules designed by legislatures to regulate markets. Despite periodic declarations of intent by political parties to lighten the burden of regulation on businesses, there is no sign of any reduction in the volume of regulation, even in a country as market-friendly as the United States. It is also noticeable that such formal rules often displace earlier informal norms. For example, as new organisations sought to participate in the London wholesale financial markets in the 1970s and early 1980s, the authorities felt obliged to replace the traditional unspoken constraints on unethical behaviour with a battery of formal regulations. It is perhaps too early to say whether this move has been successful. But it would be no doubt justified by pointing out that the new entrants to the market, mainly foreign-owned organisations, would not feel that the disapproval of other market participants was a particularly effective sanction.

Contemporary Ireland offers an interesting example of rapid economic growth in the Western world. Although the country became politically independent as long ago as 1922, as late as 1961 it could justifiably be regarded as little more than a backward agricultural region of the British economy. In that year 77 per cent of its exports went to the UK, its currency was fixed to sterling (until 1979), and external flows of labour and capital were almost exclusively directed to and from the UK. Less than 40 years later, the Irish economy is a developed industrial region of the European economy. In 1996 less than 25 per cent of its exports went to the UK, while 47 per cent went to the rest of Europe. Its GDP per capita is converging on the European average, having overtaken the UK. How has this transformation come about?

It is generally agreed that no single factor explains the sustained high growth of the Irish economy in recent years. Rather, growth has resulted from the mutual interaction of a number of factors including a changing demographic structure, an increase in human capital, the openness of the economy, investment in infrastructure, the development of a climate of social and business confidence, and some economic policies of the government. Success was also aided by two significant historical 'accidents'.[6]

What is perhaps the most interesting lesson of the Irish experience, however, comes from the trial-and-error nature of government policies. In 1932, the incoming government embarked on a trade war with Britain in which each side imposed tariffs on the other's exports. This lasted for six years. In the 1950s, there was an attempt to foster domestic industry behind tariff and regulatory barriers, a policy doomed to failure by the small size of the internal market. Between 1978 and 1982 official macroeconomic policy was so bad that it sent growth into reverse between 1981 and 1986. After 1987 the economy began to enjoy the fruits of investment in education and in infrastructure and from industrial policies (notably the attraction of inward foreign direct investment in manufacturing), which had been applied over three earlier decades. What is also significant, and supports the idea of the context-dependent nature of successful institutions, is that Irish economic policies differed markedly in some important respects from those which were being successfully pursued in the UK at the same time by the Thatcher government.

What is remarkable in reviewing the historical experience of economic growth and its implications for current policy is the apparent consensus amongst contemporary historians. North, Landes and Fukuyama all agree that, even though we do not yet know enough about the process of economic growth to be able to prescribe a formula for achieving it in the short run, any undeveloped country aspiring to do so can only hope to succeed by interpreting the lessons from the success of the advanced market economies of the Western world. Fukuyama has even gone so far as to suggest that there can be no other political objective in the modern world than a liberal democracy and no other economic objective than a market economy.[7]

The formula for economic growth offered by Adam Smith in the passage quoted on page 34 may seem simplistic, although it may not have been so wide of the mark in the late eighteenth century. Today, its greatest weakness appears to be that it takes for granted the existence of the necessary informal institutions, that is norms of behaviour. And we have seen that one cannot simply legislate to substitute formal rules for missing informal norms for several reasons. We can't observe what these missing norms might be. They are context dependent; that is they are part of a complex historical process of interaction between factors such as beliefs, institutions, organisations and events which we do not fully understand.

Furthermore, circumstances are continuously changing: it cannot be assumed that specific policies that worked in the past even in the same country will be appropriate in the future. For example, the financial market arrangements which appeared to serve Japan well for 40 years after the war no longer do so. As North observes:

> technologies change, competitive structures change, government policies change, and the way in which they operate change. If we are going to have markets that work well tomorrow, we must be continually concerned that they are going to adapt to new problems and new strategies.[8]

What does seem apparent, however, is that the account of the experience of economic growth offered by some leading contemporary economic historians is consistent with the perspective which treats economies and societies as complex adaptive systems. In both, the focus is on adaptability, and the evolution of the 'right' norms of behaviour is critical to success. In his comprehensive review of economic growth 'The Wealth and Poverty of Nations', Landes eschews any formula or even any explicit analytical framework. Yet his summing-up is implicitly congruent with a complexity approach. He writes that: 'If we learn anything from the history of economic development, it is that culture makes all the difference',[9] and that: 'Monocausal explanations will not work . . . the determinants of complex processes are invariably plural and interrelated'.[10]

For societies wishing to maintain economic growth, 'much will depend on their spirit of enterprise, their sense of identity and commitment to the common weal, their self-esteem, their ability to transmit these assets across the generations'.[11] For those poor countries wishing to achieve sustained economic growth the norms which they need to evolve are the familiar ones: 'work, thrift, honesty, patience, tenacity . . .'.[12] Again in the spirit of complexity, Landes writes: 'The one lesson that emerges is the need to

keep trying... No perfection... We must... listen and watch
well, try to clarify and define ends, the better to choose means'.[13]

What is quite clear is that equilibrium growth theory cannot
be reconciled with the historical and contemporary record.
Convergence of the growth path of economies as predicted by
the original Solow version of neoclassical growth theory (see Chapter 3) has only tended to occur amongst the advanced economies.
Persistent divergence of growth paths, on the other hand, which
appears to be the conclusion of the latest versions of that theory,
cannot explain the rise of the NICs or of China. Equilibrium
growth theory suggests an unfamiliarity with the empirical evidence and a preoccupation with the superficial aspects of the
process of growth, namely stocks of human and physical capital
and technology, while ignoring the incentives embedded in the
institutions. If the world were to operate as equilibrium theory
would have us believe, economies would reverse their direction
quickly following a radical change in relative prices. In fact, most
historical change is gradual and incremental.

11
Patterns in Economic Activity

The characteristic feature of complex adaptive systems is the spontaneous emergence of order, in the form of identifiable regularities or patterns at the aggregate level of the system as a whole, out of the apparently disorderly interactions of the individual agents or elements of the system. In Chapter 8 we noticed the emergence of a network of markets as one of the regularities of the complex adaptive system which is the market economy. In this chapter, we review four other patterns which emerge in market economies. The first two of these, the pattern of economic growth and the pattern of the business cycle are qualitative rather than quantitative. The second two, patterns in the spatial location of economic activity and patterns in production, are measurable.

The pattern of growth in the market economy

The origin of the process of growth in a market economy can be traced to that same principle of self-interest which underlies the emergence of a network of coordinated markets. Two individuals engage in a voluntary mutual exchange of goods or services because they both expect to gain from the trade. But trade between individuals leads to an eventual specialisation of tasks (or a division of labour, as Smith called it). And increasing specialisation creates the possibility of exploiting economies of scale, with the unintended but enormously beneficial consequence that greater income is available in aggregate to the group of participating individuals than would have been the case if there had been no exchange and no specialisation.

One way of appreciating the progress of specialisation over time is to perform a mental experiment. Imagine a primitive society which is composed of households engaged exclusively in subsistence agriculture. Before the process of increasing specialisation begins, each household is entirely self-sufficient; that is, there is no trade between them. Thus there is no economic system other than that which exists inside each household. Then some households begin to trade with others, exchanging goods which are surplus to their requirements as well as perhaps some labour services. As trade grows, some households will find it advantageous to specialise in the production of particular crops, others in crafts such as the making of clothes and shoes, still others in making simple tools, others in brewing, others in preparing herbal remedies and so on. Eventually some tasks will become completely specialised, in the sense that their supply moves outside all households except those which are completely devoted to their production.

At this point, a new branch of the economy, or 'industry', can be said to have been formed. After some time, this new industry will grow large enough to make profitable another task, the production of tools specifically designed for use in that industry alone. The deployment of these new tools and the reorganisation of the process of production associated with them will result in an increased productivity, the benefits of which must accrue to some members of the community. Their increased income will make itself felt in an increased demand for the goods and services produced by other households, some of which will feel able to use more specialised tools and methods in the tasks in which they have specialised. Gradually, more and more tasks will move out of households and become new 'industries'.

It is therefore possible to understand the increasing specialisation of the market economy in terms of the increased range of tasks which have moved out of the household. One hundred years ago, many households in Europe and North America still made their own clothes, brewed their own beer, made their own furniture, grew their own food, and created their own entertainment. Few do so now. But the process of specialisation continues apace.

Currently, the tasks of preparing food and looking after children are moving outside the household, and are becoming industries. For those 'households' which still are producing food, that is farms, most are largely specialised, and a whole fleet of ancillary industries devoted to the production of such goods as agricultural machinery, chemical fertilisers, and pesticides, collectively known as agribusiness, has evolved to perform for them the tasks they once performed for themselves.

If statistical evidence is sought for this pattern of growth in market economies, it can be found in the cells of input–output tables, statistical records of transactions between industries which have been produced periodically for most advanced countries since around 1950. These tables record the flows of goods and services between the different industries in the economy as well as flows for final consumption by households, government and exports. Comparisons of tables for one country for successive years will show that the number of blank or empty cells reduces over time as specialisation increases. Of course, the comparisons would be even more illuminating if such data could have been recorded over longer historical periods, say from 200 years ago for the United States. Unfortunately, the earliest table is for 1919 for that (or any) country. A series of tables for, say, Papua New Guinea, starting today and continuing at five-year intervals would reveal a similar pattern of specialisation unfolding. At the end of this chapter, we shall see how a comparison of input–output tables between countries at broadly similar levels of development reveals another interesting pattern.

We are all familiar with the idea that economies of scale can be found within a particular firm or factory: the larger the capacity of the plant the lower are the expected costs of production per unit. But what Smith had in mind was something wider: it was the increasing returns which were to be had from expanding the size of the *whole* economy (that is the entire network of markets) which was really important.[1]

In this process of increasing specialisation it is the productivity gains which arise from the use of increasingly roundabout or indirect methods of production which really matter. For example, it would be unprofitable to equip a car factory with an elaborate array of specialist machine tools if only a few hundred cars per year were to be produced. But if several car factories order the same machine tools, their costs of production will be lower and their utilisation may then be justifiable. So the decision to design, to manufacture and to use the more specialist and more productive equipment will depend on the size of the market for output as a whole, not just the market for one firm's output.

As more specialised equipment is used, this usually means that the productivity of the process incorporating that equipment rises, which in turn results in higher profits and/or higher wages. When spent, these profits and wages contribute to a further expansion of demand which in turn justifies a resort to more roundabout methods of production than those associated with the previous generation of plant and equipment. Thus a virtuous circle is established in which change is progressive and propagates itself spontaneously in a cumulative way.

The process of growth of the whole economy through increasing specialisation is both cumulative and self-perpetuating. This is undoubtedly the process which Smith had in mind when he wrote in the second part of his famous formula for growth:

> Little else is required to carry a state to the highest degree of opulence from the lowest barbarism but Peace, Easy Taxes and a Tolerable Administration of Justice: *all the rest being brought about by the natural course of things*.[2] (Emphasis added)

The market economy grows by means of successive adjustments and adaptations, and each new adjustment paves the way for another. The adoption of new technology, the accumulation of capital and the acquisition by the workforce of diverse skills are all a consequence of the overall growth of the system, not only a cause as equilibrium theory would have it. Needless to say, there is no tendency to equilibrium, a meaningless concept in this context, since different parts of the system grow at unequal rates and are subject to continuous qualitative change.

Three particular sources of increasing returns to scale can be identified, namely increased roundaboutness of the process of production, the application of new technology, and continuing organisational improvements. Of these, the first is the most important since the other two depend on it. More specialised plant and equipment is more productive in general than less specialised plant and equipment, and the extent to which it is profitable to deploy more specialised machines will depend on the current level of specialisation or degree of roundaboutness in the economy as a whole, which will in turn depend on the current size of the economy.

Two important implications of the process of growth in a market economy need to be emphasised. The first is that the potential scope for specialisation and hence for productivity gains is at any moment limited by the size of the market. This is what Smith perceived, and why he was such a passionate opponent of all restrictions on the freedom of trade, both domestic and international. He saw trade as a way by which small nations could escape from the limits to growth imposed by their size. Freedom of international trade meant that the potential rise in living standards of a country need not be constrained by the size of its home market.

It is this insight of Adam Smith which continues to inspire the contemporary movement towards the reduction of international trade barriers. In the postwar period this has been manifested in successive rounds of mutually-agreed trade liberalisation, sponsored by GATT (now the WTO). It also provides the rationale for the recent wave of regional free trade agreements such as NAFTA, the European Single Market, Mercosur and others. It is the same principle which almost certainly explains the superiority of the productivity of American industries to those of other countries. The sheer size of the US economy accounts for its superior productivity. Size also explains the exciting potential of the Russia and Chinese economies, if they can achieve and maintain political stability. A successful market economy requires a framework of legal and other institutions and above all norms of behaviour not yet fully adopted in these two countries.

As the process of specialisation has continued in the modern era, one of its consequences has been specialisation of function within industries but between countries. A typical European motor manufacturer may have some components for several models made in one country and shipped to its assembly plants in other countries. This explains the statistics of imports and exports between industrial countries which show for example both France and

Germany apparently exporting cars to each other in large volumes. Such trade flows across national boundaries appear anomalous to economists trained to believe in specialisation of production based on national comparative advantage.

Statistics of international trade also allows us to measure the importance of the process of specialisation indirectly. In the last half century, the aggregate net output of all the major countries of the world taken together has been growing at around 2 per cent per annum, while the volume of internationally traded goods has been growing much faster – at around 9 per cent per annum. Such a difference between the rate of growth of trade and the rate of growth of output can only be explained by an increasing production of specialist intermediate goods. The way in which the process of specialisation operates may also help to explain one of the features of contemporary business life which has puzzled many economists and statisticians.

The United States economy has enjoyed a period of sustained growth in aggregate output and productivity over the last seven years (to go back no further). Most observers agree that one of the ingredients of this success has been the phenomenal level of investment in computer technology by businesses across almost all industries. Yet individual firms who have made their investment in computers in anticipation of a large increase in productivity have very often found themselves disappointed. They have been unable to point to any substantial improvements in financial performance as a result of these investments.

The explanation may be that the benefits of computerisation have been dispersed over a very wide range of users (including households) who have been able to avail themselves of the services of computers and other IT facilities (which they could not otherwise have afforded), at low prices made possible by the size of the market for these goods. Of course, some part of the benefits of computerisation will have been captured by the suppliers themselves. Finally, increasing specialisation increases the number of interactions between individuals, so that growth in output and hence in specialisation is congruent with growth in complexity.

Patterns in time: the business cycle

Although growth in output, specialisation and complexity in a market economy is a cumulative and self-perpetuating process, it is far from being a smooth one. It is indeed a highly turbulent process which Schumpeter has aptly described as 'creative destruction'.

The introduction of a new technology, for example, will render unprofitable some existing process of production. Likewise, the launch of a new product on to the market will take away some of the demand from an existing product or products. In both cases, jobs are lost, and the value of some existing plant and equipment is reduced. The more specialised is the capital, the greater is the proportion of its value which is destroyed. Thus the more complex is the economy, the less likely it will be that profitable alternative uses can be found for highly specialised equipment. Although the process of growth creates more jobs in total than it destroys, the new jobs created are unlikely to be in the same place, nor are they likely to require the same skills, as those which have been lost.

The growth process is frequently disruptive at the level of the individual firm and household (as family members lose their job), but also periodically at the level of the industry and occasionally at the level of the economy as a whole. Throughout the nineteenth century, for example, there was a steady increase in specialisation amongst firms whose business was connected, directly or indirectly, to horse-drawn transport. Businesses such as coach-builders, harness-makers and the like increased both in number and in diversity. With the advent of the automobile, most of these firms collapsed, although some coach-builders were able to survive by adapting to making bodies for motor cars.[3]

These and other examples from the history of market economies suggest that growth in specialisation may be intermittent, and may even be temporarily reversible as collapses occur randomly from time to time. The important point to be emphasised is that every important advance in the reorganisation of production provokes responses elsewhere in the economy which have a further unsettling effect. These changes are not shocks which come from outside the system: they are produced within the system and by the system itself as an inescapable corollary of the nature of the process of growth in the market economy. So the process of growth is characterised by an orderly and beneficent outcome at the level of the economy as a whole, accompanied by more or less continuous disruptions at the individual level. From time to time, however, these disruptions make themselves felt at the aggregate level, and what is interesting is that such aggregate disruptions are not just random or unique, but follow a pattern to which students of the history of market economies have given a name, the business cycle. There is no recorded instance of a market economy which has experienced growth without such periodic fluctuations.

The use of the word 'cycle' suggests a mechanical regularity about these fluctuations in the level of business activity, and from time to time scholars have claimed to be able to identify cycles of regular periodicity in the rather sparse historical data which is available for a small number of countries. For example, Clément Juglar claimed he could identify the recurrence throughout the nineteenth century of a business cycle with an average duration of about seven years, attributable to periodic replacement investment in manufacturing plant and equipment. Likewise, a shorter cycle of two to three years based on an alleged rhythm in inventory fluctuations was claimed by Joseph Kitchin. The Russian economist Nikolai Kondratiev (1892–1931)[4] thought that he had identified 'long waves' of economic activity recurring every half-century or so which he explained by the renewal of long-lived infrastructure investments like those in canals and railways.

Given the inadequacy of the relevant data before 1900, together with conflicting indicators between and within countries since then, claims for the existence of any cycles with regular periods are difficult to sustain. So far as the UK is concerned, there seems to be agreement amongst scholars on the existence of only four major cycles. These were the downswing following the Napoleonic Wars (1815–48), the 'Great Depression' (1873–96), the second Great Depression (1929–34) and the Recession of the 1970s (1973–8). In the USA, careful research by the NBER on postwar turning points in the US economy show conclusively that recent business cycles do not occur with regular frequency.[5]

Although fluctuations in economic activity appear to be irregular in size as well as in timing, and although they evidently differ in detail from one historical circumstance to another, nevertheless they do tend to exhibit a common pattern which is not quantitative but qualitative.

The elements whose interaction accounts for the cyclical character of business activity can be divided into four groups. They are (a) real factors, such as flows of savings and investment; (b)monetary factors, such as the supply of credit and the rate of interest; (c) human factors, not only fear and greed but bouts of unwarranted optimism and pessimism, changing attitudes to risk, and periods of herd-like behaviour; and finally (d) there are government policies, in particular intervention in markets by means of monetary and fiscal policies.

The most important real factors in the business cycle are innovations. Innovations are simply new ways of doing things which result in investment: not only the launch of new products, but new processes, new forms of organisation and new methods of distribution. The principal reason for the undulating pattern of the business cycle is that innovations tend to cluster together in time, partly because of the nature of breakthroughs in technology, but largely because when one firm thinks it sees a new business opportunity other firms follow. Recent clusters of innovation-led investments can be observed in circumstances as diverse as the supply of retail financial services by telephone in the UK (late 1980s), the trading of derivative financial instruments in the USA (mid-1990s), the supply of semi-conductors to a global market (mid-1990s), and in telecommunications (1998–9).

The upswing of the cycle typically starts off with a rise in investment associated with an innovation. This is usually accompanied by an expansion of credit and growing business optimism. With producers vying with each other for assets and the new output not yet on the market, the result of the expansion of credit is inflationary: asset prices are bid up, the mood of optimism is reinforced and the boom is underway.

The innovation attracts a host of imitators. Eventually, it becomes clear that there is going to be excess capacity and that asset and product prices will fall. The downswing begins. Falling prices squeeze profits and put downward pressure on wages. Some of the innovators, perhaps the later ones, and their creditors will be left with losses. If wages are 'sticky' in moving downwards, the unemployment caused by old plant being displaced by new will be aggravated. The squeeze on profits leads to a collapse in confidence. A depressed state of business lasts until new investment sets off a further upswing.

In each phase of the cycle, the monetary and human factors tend to reinforce the real factors; that is, there are positive feedback loops. In the upswing, banks are willing lenders, but in the downswing when credit is most needed by those businesses which are solvent in the long term but temporarily short of liquidity, banking principles and banking regulations combine to limit the supply of credit. The human factors are equally destabilising: in the upswing, there is unwarranted optimism, and in the downswing unwarranted pessimism.

Sometimes, at the bottom of the cycle, the human factors can be more powerful than the real and monetary ones, and may even overcome countervailing government policies. In a depression, businessmen may not be persuaded to invest no matter how low are the rates of interest, and consumers will save every penny that comes their way. In such cases, an economy may languish in stagnation for some time, as recently in Japan. But this is exceptional. In most circumstances, market economies will find their own way out of recessions without too long a delay.

Most people became aware of the series of financial crises in East Asia which began in Thailand in the summer of 1997 and spread to many other countries. Few made the connection between these financial crises and the preceding upswing in real economic activity which had been in motion throughout much of East Asia for at least a decade, part of what is sometimes known as the 'Asian miracle'. The relationship between financial crises and the business cycle is described below, but it is perhaps appropriate to say something first about the mechanisms by which both cycles and crises may be transmitted from one country to another.

So far as real factors are concerned, it is not difficult to see how variations in business activity can be transmitted from one economy to another if the two are linked by trade or by flows of direct capital investment. Although the market for financial capital is a global one, it is less obvious that monetary disturbances and mood swings in one country should necessarily lead to similar upheavals in another. Those who argue in favour of 'contagion' point to the example of the 1930s, where a succession of financial crises forced one country after another off the gold standard. Today, it is alleged that a similar succession of crises has compelled one country after another to give up their fixed exchange rates with the dollar. But it may well be that in both periods the difficulty lay in official attempts to defend unsustainable fixed rates of exchange between individual national currencies and gold or the dollar rather than in the international transmission of purely human factors such as fear or greed.

Conclusions on business cycles

Business cycles are not the result of external shocks inflicted on the economic system, although they may be triggered off by them. They are an inescapable part of the process of growth which depends fundamentally on a sequence of innovations that are clustered in time and therefore disruptive. The expansionary impact of an upsurge in real investment is reinforced by monetary and human psychological factors. The subsequent recession has a remedial function in that it eliminates those investments which should not have been made. Since no-one can tell in advance which ones these are, this means that a market economy grows by trial and error. To mix two popular metaphors, just as the rising tide of the cyclical upswing lifts all boats, so the falling tide of the downswing reveals who is naked. It follows that the idea of smooth, sustainable growth in the market economy must remain a politician's pipe-dream until such time as innovations cease being clustered in time, and there are no more fluctuations in the supply of credit and no more collective mood swings from optimism to pessimism and back again.

Financial crises

The downswing phase of the business cycle is often ushered in by a financial crisis. Such crises tend to be more dramatic than business cycles, and are thus more widely reported and remembered. Almost everyone has learned that the Great Depression of the 1930s in the United States was precipitated by the collapse of prices on the New York Stock Exchange in October 1929, 'The Great Crash'. But the Wall Street Crash of 1929 was not the first financial crisis to have impressed itself on popular memory. The 'South Sea Bubble' of 1720 (speculation in London in shares of the South Sea Company and in Government debt) is still remembered for what Adam Smith called 'the folly and negligence' of the Company's management, and 'the knavery and extravagance' of their stock-jobbing operations.[6] Some economic historians believe that this episode marked the beginning of a downswing in economic activity in England which effectively delayed the onset of the Industrial Revolution there.

Kindleberger has identified the occurrence of more than 30 major financial crises in various countries between 1720 and 1990. Each of these differed from one another in various particulars. They were triggered by different events, the objects of speculation on each occasion were different, and the expansion of credit took different forms. Nevertheless, a common pattern in the phases of each financial crisis is discernible.[7]

There is first of all an external event which triggers off the process, followed by a phase of speculation or 'overtrading', defined as the purchase of assets for resale rather than for use or for income. This phase is invariably marked by widespread examples of irrational behaviour by individuals, perhaps in part explained by the observation that 'there is nothing so disturbing to one's well-being and judgement as to see a friend get rich'.[8] Accompanying this speculative phase is a period of monetary expansion in which credit is extended to speculators by banks and other institutions. Sometimes this is unnecessary. On the Kuwaiti Stock Exchange between 1977 and 1982 shares were bought with post-dated cheques. By the time of the bust, worthless cheques amounting to some US$91 billion had accumulated.[9]

Eventually, when expectations of profit are overtaken by fear of loss, there is a flight from whatever assets have been the subject of speculation into more liquid assets. This phase Kindleberger calls 'revulsion'. Revulsion leads banks to stop lending on the security of the assets which were the object of the speculation.

This may lead to a phase of panic which, like the earlier speculation, feeds on itself. Confidence will not be restored until one
of three things happens: (a) trading is discontinued, (for example,
when an asset has fallen in price by more than a given percentage
in a single session – the so-called circuit-breaker device), (b) prices
of the assets fall so low that people are tempted to buy them
once again, or (c) a lender of last resort succeeds in convincing
the market that money will be made available in sufficient volume
to meet the demand for liquidity.

So, like the business cycles of which they often form a part,
financial crises exhibit a regularity of pattern in their occurrence
as well as a considerable diversity of detail.

Patterns in space: the size distribution of cities

The patterns in economic activity we have looked at so far, the
network of markets, the patterns in growth and in cycles are all
widely recognised to exist. But in none of these cases is the pattern
really clear-cut or distinct in a quantitative sense. In the case of
our next subject, the size distribution of cities, the regularity of
the pattern is almost too good to be true.[10]

Krugman ranked the 130 cities of the United States in order of
their population size. If the rank of each city is plotted against
its size on a graph using a logarithmic scale, the result is close
to a straight line with a slope of almost exactly minus one. This
result is sometimes known as the rank-size rule: the size of a
city is inversely proportional to its rank. If the rule held exactly,
the third largest city in the country would have a population
one-third of the size of the largest one, and so on.

In fact, so far as the contemporary United States is concerned,
the rule does not hold exactly at the top end of the scale: Los
Angeles is considerably more than half the size of New York.
But further down the ranking the fit is astonishingly close. For
example, the 100th ranked city (Spokane, Washington), has 370 000
people, while the tenth ranked city (Houston, Texas) has 3.85
million people. Data from the past show that this regularity has
held true in the United States for at least one hundred years.
However, without some modification, it is less strong in other
countries.

The regularity of the pattern of the size distribution of cities is unusual amongst economic phenomena, but it is unmistakable. Its origins are less clear. A number of theories have been put forward to explain its occurrence, but none of them are entirely satisfactory. Perhaps the most interesting is that of Herbert Simon, who suggested that the growth of a city depends on the number of people directly or indirectly employed by an entrepreneur.[11] If new entrepreneurs are usually inspired by existing enterprises, so that any given city will generate new ideas in proportion to its population, and if only a small fraction of entrepreneurs move far from the source of their ideas, then this could possibly explain the phenomenon.

Equally simple, and equally hard to explain, relationships between size and frequency appear in some non-human complex systems. For example, amongst physical phenomena there is a relationship (known as the Gutenberg–Richter Law) between the size of earthquakes and the frequency of their occurrence. The frequency with which meteorites exceeding any given diameter hit the Earth follows the same kind of rank-size rule. Amongst biological phenomena, the number of animal species that exceed a given size follows a similar pattern. What could such diverse phenomena have in common?

Krugman suggests that a rank-size relationship emerges when three criteria are satisfied. First, the objects of study must be subject to significant growth over time. Second, the growth rate of any individual object must be random, so that over time you get a wide range of different sizes. Third, large objects must grow on average neither faster nor slower than small ones. This behaviour suggests a different type of self-organisation. Whereas all the patterns of economic behaviour we have hitherto discussed represent examples of order emerging spontaneously from circumstances of apparently unstable disorder, the size distribution of cities and other rank-size patterns illustrate the emergence of order from random growth.

Another example of the emergence of order from random growth is worth mentioning, although it has nothing to do with economic activity, because it is a phenomenon which exhibits a regularity of pattern even more extraordinary than that found in the size distribution of cities. Suppose a ceramic object such as an urn is thrown hard against a wall so that it smashes to pieces. It would be hard to imagine an act which so vividly embodies the destruction of order. And yet, if the shattered fragments of the urn are carefully collected, counted and weighed, it will be found that the pieces obey a simple rank-size rule: the frequency of the fragments in each size class will be inversely proportional to their weight.

An even more remarkable property is to be found in the frequency distribution of the shattered fragments of the ceramic object. Another way of expressing a rank-size relationship is to say that the frequency distribution of object sizes follows a power law. In the present case, the number of fragments, N, whose size exceeds some given weight, W, is proportional to W raised to some negative power (exponent), p, that is $N = W^{-p}$. If the ceramic object were a Grecian urn then the exponent p turns out to have a particular value, whereas if it were a ceramic sphere it would take on a different value. In other words, the apparently complex and disorderly process of fragmentation (which can be regarded as random growth with a negative sign), produces a result which follows the simple regularity of a rank-size rule or power law, and the value of the exponent of that power law identifies the particular shape of the original object.

Patterns in production

There is another important regularity to be found in contemporary market economies. It is one which may be even less familiar to economists than the pattern of the size distribution of cities, but like that pattern it is a measurable one. This regularity concerns the pattern of transactions between the different sectors of production in a market economy. At first sight, there is no similarity between the patterns of transactions to be found in the statistical accounts of different countries, but when the more important transactions are isolated a distinctive common pattern presents itself.

Since the Second World War, the governments of most OECD countries have, as part of their national accounting procedures, periodically prepared statistical tables showing the value of the annual flow of goods or services from one sector of the national economy to another. These input–output tables, as they are called, arrange the different sectors in the same order in both rows and columns. Each cell of the input–output table shows the value of raw materials, semi-finished goods and services delivered from the sector of production corresponding to the row of the table to the sector corresponding to the column. These inter-industry flows are said to represent the structure of production of the economy concerned. By arranging the sectors of two input–output tables, each representing a different country, in a similar order it is possible to make a comparison of the structure of production of the two countries concerned. In order to eliminate differences in scale, it is usual to form coefficients of proportionality by dividing the value of the flow in each cell by the corresponding column total. Even after this adjustment, however, no common pattern can in general be detected between such tables.

However, when the smaller coefficients are removed, eliminating the detailed peculiarities as well as the minor results of differences in accounting conventions between countries, a distinctive pattern of inter-industry relationships emerges. This pattern seems to be shared by many countries at a broadly similar level of economic development. Moreover, on closer examination the pattern is not just a statistical one, but seems to be capable of a plausible economic interpretation.[12]

A visual inspection of the tables shows that this common pattern, or fundamental structure of production, is composed of four groups or blocs of industries sharing certain easily recognisable characteristics. The blocs are composed of Metal-working industries, Non-Metal-working industries, Energy industries and Service industries. Furthermore, the pattern shows that these blocs are almost self-contained. This implies that, despite the enormous progress in technology, combination between metals and non-metals in the production process has been quite limited. Although it is easy to think of specific examples to the contrary, the tables show that their practical importance is overshadowed by inputs of physically similar materials. Whether the tendency for metals and non-metals to combine with each other increases or diminishes over time remains to be established. It might be argued that, at an early stage of development, metals had necessarily to be combined with other materials. On the other hand, it might be supposed that the progress of technology would open up possibilities of physical combinations which did not exist before.[13]

A second important characteristic of the common pattern is its triangularity. In the present context, triangularity means that output at one stage of the process of production is not used as an input at an earlier or higher stage. Clearly, in the longer term the economic relationships of a modern economy are circular: produced capital goods themselves become inputs in the process of production. (The input–output tables record only flows of current or intermediate goods.) Even confining ourselves to these current flows of materials and intermediate goods and services between industries, the full tables show sufficient circularity that they can seldom be satisfactorily triangulated. But once the minor transactions are removed, the underlying common pattern is seen to be triangular in form. In fact, it is triangular in two senses. Not only is it (almost) triangular in all its elements, but the blocs follow a triangular order with respect to one another.

The existence of this easily identifiable pattern underlying the structure of production of different economies in different years, is yet another example of that phenomenon of regularity of pattern but not of detail characteristic of advanced market economies and of complex adaptive systems generally.

12
Adaptation in the Market Economy

Perhaps the most fundamental distinction between mechanical systems and complex adaptive systems concerns the role of knowledge or information. In the former, complete information about the future as well as the present is universally available at all times, so there is no need for adaptation or adjustment of any kind. In the latter, information is acquired and acted upon as the system evolves. This method of adaptation may be described as trial and error or learning by doing.

All agents in complex adaptive systems must adapt eventually if they are to survive. In the human complex adaptive system which is the market economy, some agents are more important than others in performing the function of adjustment, more important that is from the point of view of the development of the system as a whole. These people have been given the distinctive name of entrepreneurs. Entrepreneurship is a distinctive phenomenon of human systems: no corresponding phenomenon exists in non-human complex adaptive systems. Such an important function requires an appropriate explanatory theory. The idea of the entrepreneur as hero of the market economy can be traced to the work of the Austrian economist Joseph Schumpeter, who first put forward a theory of entrepreneurship as long ago as 1912.[1]

What exactly is an entrepreneur? Schumpeter's definition was that an entrepreneur is a businessman who innovates, that is who tries a new way of doing things, whether that is launching a new product, introducing a new process of production or method of distribution, or opening up a new market. This behaviour is to be contrasted with the type of manager who predominantly carries out routine functions within a more or less unchanging business environment, keeping the business ticking over in a fairly repetitive process. So Richard Branson is an entrepreneur, while the chief executive of a water supply company is probably not. According to Schumpeter, entrepreneurship is a function that requires peculiar aptitudes which are not evenly distributed throughout the population.

Other economists who have written about entrepreneurship have emphasised different aspects of the function. While some, like Schumpeter, would reserve the phrase for a very small number of people who create significant market innovations, others, like Mises, believe that all human action, even as simple a task as crossing a busy road, has an entrepreneurial character.[2] What these different perspectives have in common is the idea that entrepreneurial behaviour is adaptive behaviour. Entrepreneurs operate by exploiting information which they acquire as they go along. They are always on the lookout for new profit opportunities, buying cheap and selling dear. But entrepreneurship is not a deliberate act of search or of learning: it is the largely chance discovery of existing opportunities which neither the entrepreneur nor any of his or her competitors had perceived before. In short, to be a successful agent of adaptation in a market economy, one must be able to perceive opportunities for possible profit which others have not recognised, and then mobilise the resources necessary to exploit them.

Today, almost everyone in the Western world has accepted the proposition that the entrepreneur is one of the key elements in the successful performance of a market economy. This was the message that Mrs Thatcher brought to a sceptical British public following her General Election victory in 1979. It was a message that reverberated around the world, and eventually won over her political opponents. Now even social democratic governments vie with each other in proclaiming the importance that they attach to entrepreneurship as an ingredient of national economic performance. Despite the practical importance of entrepreneurship to the success of the modern market economy, the astonishing fact is that there is no place at all for that function in equilibrium theory. Nearly all those economists who have written about entrepreneurship belong to the Austrian School of economic thought. Why has neoclassical economic theory ignored the entrepreneur?

It is not just because it would be impossible to squeeze such an unquantifiable concept into the mathematical models which equilibrium economics believes to be the hallmark of scientific method. It is primarily because equilibrium theory assumes that businessmen and households operate in a world of perfect information; that is, they have at all times complete knowledge about all the options open to them and the consequences of choosing between them. All the decision-maker has to do is to compute that combination of options which gives the best results (the optimum) in terms of his or her desired objectives. Businessmen are assumed to select that combination of investment opportunities which will maximise their profits. Households are assumed to purchase that bundle of consumer goods which maximises their satisfaction. In a world of perfect knowledge of the present, and perfect certainty about the options available in the future, there is no need for an entrepreneur. A calculator is all that is necessary.

But of course the world of business and economic activity is not like this. Far from there being perfect knowledge, there is radical ignorance. Mostly, people do not even know what it is that they do not know. Decisions and actions have to be taken in an environment where knowledge is dispersed; that is, different people know different things, and what each knows is often tacit, fragmentary and inaccurate. In these circumstances, optimisation is simply not possible. Contrary to the assumptions of equilibrium theory, human beings do not respond to their continuously changing business environment by making 'rational' (that is consistent) decisions in pursuit of well-defined objectives on the basis of complete information. The outcomes of even the limited choices of which they are aware will occur in the future, and the future, because of the complexity of human interactions and the influence of chance, is at best uncertain, at worst unknowable.

Recognising the limitations of the assumption of perfect information, equilibrium theory has responded by introducing a slight modification called search theory. Here, it is admitted that typically some information is missing, but it is assumed that those whose information about a market is incomplete know what is the information that is missing, what its precise value is to them, and how and at what cost it is possible to obtain it. This is like saying that a businessman need only commission some market research in order to learn with certainty everything he needs to know about the present and the future commercial opportunities open to him including the profits he could make on each possible venture.

In practice, no matter how extensive the market research, the problem remains that the future cannot be known with certainty. No matter how good the market research, the success of a new product will never be known prior to its launch. In general, the customer does not know what he or she wants until s/he has tried it. Hence the widespread use of pilot schemes and field trials to discover customer reactions to a new product. As a result of these trials some detail of the product or its method of distribution may often be altered. In other words, the supplier frequently adapts to his market environment through a process of trial and error.

In these circumstances of fundamental uncertainty the entrepreneurial function becomes essential to the successful functioning and progress of a market economy. Discovery of the optimal course of action through rational calculation is simply not possible. The only way to reduce uncertainty is by testing options in practice. As a result, new information is discovered, including information about new options which become available, and then exploited. This is the defining characteristic of entrepreneurship. It is also the defining characteristic of adaptation in complex systems. In the static terminology of equilibrium theory, entrepreneurs exploit imperfections of existing information. In the evolutionary language of complexity theory, they acquire and exploit new information as the system evolves.

But entrepreneurs do not just adapt blindly, seeking out opportunities for short-term gain in order to survive. There is an additional distinctly human element in entrepreneurship which further distinguishes it from adaptive behaviour in non-human systems. Individual entrepreneurs often have personal goals, and make and execute plans to achieve them. It should not be assumed that their behaviour is always characterised by consistency in pursuit of their objectives or that these objectives are exclusively financial. Many, if not most, entrepreneurs get satisfaction from the creative aspects of their function. More than sixty years ago Keynes reminded us of the important role of what he called 'animal spirits' in investor behaviour.

Learning

Entrepreneurship, adaptation and progress through trial and error are terms which are broadly synonymous. They can all be summarised in the phrase 'learning-by-doing'. In biological systems, the more sophisticated an organism is, the greater is the role played by individual learning compared to the in-built schematics programmed by evolution.[3] Human beings are capable of learning very much faster than other, non-human, organisms.

There are broadly two ways for human beings to learn, through formal instruction or by learning from experience. The first method is peculiar to human beings, and appears to be by far the fastest and most effective way of learning. It has enjoyed explosive growth throughout the Western world, as illustrated by the growing proportion of the population in the advanced economies who go through some form of post-school education.

So far as learning about the market economy is concerned there has been a dramatic expansion of enrolment in business schools since the war, and the number of MBA programmes on offer throughout the world continues to increase. Yet not all businesspeople are convinced that this is more than a social fashion, and that the best way to learn how to succeed in business is through attendance at formal instruction rather than through experience. Amongst the sceptical is the enormously successful American financier, Jim Rogers. He is of the view that a young person wishing to embark on a business career will learn more from the experience of attempting to start a business (and more often than not failing) than from attending classes at a business school. Consequently, he advises such a person with US$100 000 to invest to spend it on a business start-up rather than on the costs of attending business school.[4]

Oscar Wilde was also on the side of the sceptics. He wrote that: 'education is all very well, but nothing that is really worth knowing can be taught'. While this proposition is clearly not true of the natural sciences and their derivative technologies, it may well be true for those large areas of human behaviour including business and economic activity where knowledge is necessarily incomplete. In these circumstances, trial and error or learning by doing may often be the best way of making progress or even just surviving.

Adjustment and disruption

Entrepreneurs in the market economy are not just passive respondents to change: it is they who initiate it. They are the people who are responsible for introducing new products, new processes and new methods of distribution into the marketplace. They thus confer enormous benefits on society as a whole. At the same time, these innovations cause great disruption: the new product displaces an existing one, the new process renders an older one unprofitable, and new methods of distribution likewise make current methods no longer viable. If people could move without difficulty from old to new jobs, and if capital could also move from one task to another, there would be no problem of adjustment. But the new jobs may be in a different location, often a different country, from those destroyed, while a machine which has been built to perform one task cannot normally be adapted to perform another. In a primitive subsistence economy, simple tools can be switched from one task to another. In the modern market economy, capital equipment has become so specialised that this is no longer possible. Change inevitably means disruption and as the rate of change increases, so the level of disturbance will rise, too.

The social cost of such disturbances inspired nineteenth-century critics of the market economy such as Marx, Zola and Upton Sinclair. Today, the remarkable coincidence of worldwide deregulation of markets together with rapid rates of change in technologies has brought a renewed sense of insecurity to workers in the Western world. Sometimes they respond by demanding from their political representatives that they should provide new job opportunities to replace the old ones which have been destroyed. More often they seek protection against the market forces which threaten their jobs. The problem is that the same forces which bring rising living standards, for example imports of cheap electrical goods from Asia that replace the more expensive domestic sources of supply, are also the ones which are responsible for destroying domestic jobs. One cannot have one without the other: disruption and growth in the market economy are two sides of the same coin. The task for politicians is to make the adjustment process as painless as possible, while resisting the pleas for protection, which, if granted, would arrest the growth in the living standards of society as a whole.

We have in this process of endless adjustment the very essence of how market economies work. Yet, because of its assumption of instantaneous and perfect adjustment in all markets simultaneously, equilibrium theory has nothing to say about the process at all. As Greider observes ironically:

> neo-classical economists always leave out the fun part of adjustment, in which millions of people lose their livelihood, countries lose chunks of their industrial base, and firms go belly-up or are forced to merge with other firms. They use the word adjustment as a euphemism to brush this process aside, but most people, and most firms, regard that process as what matters most.[5]

While critics of the market economy focus on the disruptive effects of competition, its defenders focus almost exclusively on its beneficial effects. Because the two sides share no common theoretical ground, any meaningful discourse and hence any progress in mutual understanding is almost impossible. Each side is reduced to assembling lists of facts in support of its assertions. Although Will Hutton's recent critique of the contemporary market economy in Britain[6] achieved bestseller status, it failed to set off a serious debate simply because of the absence of any commonly accepted theory about how the market economy works. Since equilibrium theory has nothing to say about how a market economy really works, nor what accounts for its successes and its failures it cannot be used to rebut Hutton and similar critics. For their part, such critics seem to lack any coherent theory of how the market economy works.

Adam Smith, the founding father of modern Economics, was not only an admirer of the market economy, but he wanted to understand how it worked. He saw how the individual efforts of millions of workers were coordinated without any central control or conscious cooperation to satisfy the needs of society as a whole. He also observed that this task was performed by a network of markets, but he could not explain how coordination was brought about; he was obliged to fall back on the metaphor of the Invisible Hand. When we perceive the economy as a human complex adaptive system, we can begin to understand how it delivers a tolerable degree of coordination and growth, while at the same time creating and eventually resolving problems of adjustment.

Patterns of corporate organisation

The idea of an entrepreneur suggests a business under the control, if not the ownership of a single person. But this has long ceased to be the typical form of corporate organisation, if it ever was. What relevance does the concept of entrepreneurship have in the contemporary market economy where the large corporation is the dominant form of organisation?

Before the Industrial Revolution which began towards the end of the eighteenth century, commercial organisations, whether they were merchants or manufacturers, were typically organised as partnerships or else they were controlled by one person who combined the functions of ownership and management. As developments in the technology of steam power in the nineteenth century made scale in manufacturing more attractive, there was a growing tendency toward separation of ownership from management. At the same time, there was an increasing trend toward the amalgamation of small specialist companies into larger vertically-integrated multifunctional companies under the control of a single general manager. The introduction of the company with limited liability as a permitted form of corporate organisation made it less risky for owners to allow managers to act in an entrepreneurial fashion.

In order to successfully exercise his entrepreneurial functions, the manager would have had to possess a detailed knowledge of the whole range of the firm's operations, and of the several markets in which it was engaged. But by the beginning of the twentieth century, many companies had grown too large in terms of their own internal resources and were facing too complex an external business environment for a single person to be able satisfactorily to exercise entrepreneurial responsibility for the whole company. Consequently, after the First World War a new form of corporate organisation began to emerge which allowed for an increase in the number of people within a company who could exercise such responsibility.

This was the multi-divisional corporation, where each division was headed by a general manager to whom some strategic as well as operational responsibilities could be delegated. The chief executive of the company remained in overall control, but there was now a layer of divisional general managers between himself and the operations of the company. The invention of this form of corporate organisation is widely attributed to Alfred P. Sloan, who was Chief Executive of General Motors from 1923 to 1945. This pattern of organisation was widely and successfully adopted over the next sixty years, particularly inside the United States. It made possible the emergence of the global multinational corporations with which we have become familiar, and it provided the opportunity for the growth of management as a distinct profession.

Today, however, it is being suggested that this corporate form is itself becoming outdated. The external environment facing even the largest business has become markedly more complex than it was in the 1920s, and the rate of change in the business environment has accelerated. It is suggested that the level of complexity now outstrips the entrepreneurial capacity of the existing form of corporate organisation.[7] A new form of organisation has therefore been proposed.

The new form would expand still further the number of people within the corporation who could exercise entrepreneurial responsibility. It would do so by dividing the organisation into scores or even hundreds of smaller business units, and devolving a degree of authority to the heads of these 'little companies'.

Does such an organisational form actually exist anywhere? At the present time, the best-known example may be provided by the Swiss–Swedish engineering company ABB. For this company, its Chairman, Percy Barnevik, has performed the pioneering role performed for General Motors over seventy years ago by Alfred P. Sloan. It remains to be seen whether the new organisational form will be adopted as widely as was its predecessor.

When we look at the sequence of dominant forms of corporate organisation in the most advanced market economies over the last two hundred years, we see each one emerging as a modification of an earlier form. To what extent did each of these patterns of corporate behaviour emerge spontaneously, and to what extent were they the result of conscious human planning? In the case of the legislation which enacted the principle of limited liability, one can identify an element of deliberate design at the level of society as a whole. In the other cases, adaptation seems to have occurred at the level of the individual firm, and then spread by imitation.

13
Implications for Economics

We can begin by restating the essential differences between the approach to the study of economic phenomena represented by neoclassical equilibrium economics, and by what we may call the complexity approach.[1]

In the former approach, the problem is to examine the properties of certain overall equilibrium states of the economy which are derived from the rational choices of individual agents who are optimisers. In general equilibrium analysis, the 'states' are sets of prices, while in game theory they are sets of strategy assignments with associated payoffs. But this approach can tell us nothing about the process by which these states of equilibrium come into existence, nor can it describe how the state of the economy changes over time.

In the dynamic variant of the equilibrium approach, a system of difference or differential equations describes how a set of variables representing the state of the economy changes over time, and the resulting trajectories are investigated. However, this is really a 'moving equilibrium' condition, which has nothing to say about the interactions between individual agents and their relation to aggregate outcomes.

Neither of these approaches can accommodate the appearance of new variables, let alone new entities, new patterns or new structures. Indeed, all those economic variables which are not readily quantifiable as well as all non-economic variables are omitted from the beginning. Perhaps the fundamental difference between the two approaches concerns the problem of knowledge.

In the equilibrium approach, all agents are assumed at the out-
set to be in possession of a stock of common knowledge. Any
uncertainty is evaluated probabilistically, and a course of action
is chosen which maximises expected utility.

In the complexity approach, no such common knowledge normally exists. Each agent interprets the world according to their
own experience, a world which is complicated by the presence
and actions of other agents, and which is therefore always changing. Any common knowledge which agents have about one another
must be attained through specified cognitive processes operating
on experiences obtained through concrete interactions. Common
knowledge cannot be assumed into existence. It follows that while
agents may seek to do their best in any given situation, there is
no question of optimising behaviour in the sense of equilibrium
theory. This is not because agents lack the necessary information processing capability, but because the concept of an optimal
course of action cannot even be defined.

According to the equilibrium approach, agents do not interact
with one another directly, but only through impersonal markets. This is another conceptual unreality, often overlooked. Not
only consumers, but also producers, continuously imitate each
others' behaviour, and this cannot be dismissed as a peripheral
phenomenon in a modern economy. Indeed, it may be said that
when one supplier introduces an improved product or service,
or even a superior method of organisation into a market, then
imitation is the most common way in which it is diffused. It is
true that in game theory players interact with one another, but
interaction patterns are usually simple and extreme. Except in
transactions cost economics, where a simple hierarchical structure is supposed to obtain, the internal structure of the agents is
abstracted from.

On the other hand, in the complexity approach all action involves interactions among agents. Economic functions are both
constrained and performed through recurring patterns of interactions between agents ('networks'). Such interactions are shaped
by rules and norms, that is, by institutions. And economic entities (firms industries, national economies) have a recursive structure;
that is, they are themselves composed of other entities. But the
structure of entities in this sense is not necessarily hierarchical.

Component entities may be part of more than one higher-level entity, and entities at multiple levels may interact. And they can give rise to new entities and new patterns of interaction at both higher and lower levels.

The complexity approach focuses attention on processes and on emergent structures, transient phenomena, as much as on outcomes. In particular, it wants to understand how new goods, new technologies, new theories and new organisations come into existence. In other words, it wants to know how the economy really works. In the world of perpetual novelty which characterises the real economy and which is also represented in a complex adaptive system, outcomes do not correspond to equilibrium states. The complexity approach can be summarised as 'seeking emergent structures arising in interaction processes, in which the interacting entities anticipate the future through cognitive procedures that themselves involve interactions taking place in multilevel structures'.[2]

Hitherto, complexity theorists have explored a number of specific economic problems which they have identified as being amenable to their approach, and have tried to develop methods for relating theories based on their principles to observable phenomena. Although a wide range of topics has been addressed, and they are modest about their successes, it seems as if they have underestimated the scope of applicability of the complexity approach so far as economics is concerned. This may be because the approach has been understood almost as if it were a 'transplant' from the natural sciences. It has been presented very largely as if its success depended upon it being a novel mathematical technique, rather like linear programming when it was first introduced in to economics in the 1950s.

In fact, even a cursory knowledge of the history of economic thought will confirm that the essential principles of complexity theory, that is the concepts of self-organisation and evolution, and even the concept of complexity itself can all be found in earlier economic literature. Economists in the classical as well as the Austrian tradition anticipated natural scientists in forming these concepts. One might speculate that if the teaching of the History of Economic Thought had not been driven out of the economics curriculum in the postwar enthusiasm for

equilibrium economics, there might have been a better understanding and faster acceptance of the complexity approach amongst
academic economists.

As we have seen, the notion of self-organisation is implicit in
Adam Smith's account of the coordinating function performed
by markets, and is quite explicitly treated by Hayek.[3] Hayek also
discussed at length the concept of complexity in his 1961 paper
'The Theory of Complex Phenomena', and explored that issue
in his writings on methodology twenty years earlier.[4]

So far as the concept of Evolution is concerned, it appears
that Darwin got the idea for his theory from reading Malthus.
Hayek remarks that in the nineteenth century no social scientist
'worth his salt' had anything to learn from the natural sciences
about Evolution. In support of this contention, he cites a series
of works going back to the seventeenth century.[5] The idea of
the 'spontaneous' emergence of human institutions, an idea to
which Menger attached so much importance, can be found in
Ferguson and Mandeville as well as in Smith. It is to Ferguson
that we owe the famous phrase about institutions being 'the result
of human action but not the execution of any human design'.[6]

Economists in the Austrian tradition share a common emphasis on the dynamic nature of economic activity, on the emergence
of institutions, the imperfect nature of knowledge and the inte-
gration of political and social issues with economic issues beyond
market exchange and production. So far as they are concerned,
the essential function of economics is how to explain observed
phenomena in terms of purposive human action, and how to
trace the unintended consequences of these actions.[7] To address
these questions they generally subscribe to three methodological
principles: individualism, subjectivism, and a focus on processses
rather than equilibrium states. When the term 'individualism' is
replaced by the more contemporary phrase 'agent-based', and
when subjectivism is extended from value formation to include
the formation of expectations,[8] then it is not difficult to see the
complexity approach to economics as being congruent with the
Austrian approach. It then becomes possible to think of the complexity approach as being simply the latest manifestation of an

unbroken tradition in classical economic thought from Smith, Hume and Ferguson through Malthus, Mill and Menger to Marshall, Schumpeter and Hayek.

From this point of view, the development of neoclassical equilibrium economics after Marshall, with its preoccupation with questions of value rather than growth, its static methodology, its assumptions of optimising behaviour and perfect knowledge and its exclusion of institutions, entrepreneurship and all other human aspects of economic behaviour, appears as an intellectual cul-de-sac, a branch from the main line which has come to a dead end. It is, in Lakatos' (1970) phrase a 'degenerate research programme'.

One qualification must be made to this judgement. The application of the methods of classical mechanics to economics at the elementary level has been enormously helpful. The simple apparatus of partial equilibrium analysis (Marshall's cross) is an extremely powerful device for illuminating the issues which arise in the interaction of supply and demand in particular markets. Likewise, Walras' formulation of the conditions for a general equilibrium of markets also has an important pedagogic value in introducing beginners to an understanding of the principle of the interdependence of markets. The great mistake which has been made has been in the unwarranted extension and overrefinement of these methods, which has brought the discipline of economics to its present impasse. It was an extension which Marshall explicitly warned against, but his warnings were ignored.[9]

The limitations of mathematics in economics

The biggest difference between Classical and Austrian economists on the one hand and 'complexity economists' on the other seems to be in the use of mathematics. Most papers in complexity theory use a mathematical exposition. Since the essence of complexity is non-linear interactions, then the mathematics is non-linear. This means that the linear algebra used by neoclassical economists, which exploits fixed points and systems of differential equations, is of no use in understanding complex adaptive systems.

What is required for complexity theory, it is said, are 'new classes' of combinatorial mathematics and population level stochastic processes, in conjunction with computer modelling.[10] These mathematical techniques emphasise 'the *discovery* of structure, and the *processes* through which structure *emerges* across different levels of organisation'.[11] A new discipline, 'agent-based computational economics', has staked out a territory. It defines its boundaries as the 'computational study of economies modelled as evolving decentralised systems of autonomous interacting agents'.[12]

Marshall, as we have seen, was wary of the use of mathematics in economics, and most Austrians have been openly sceptical. It may be significant that Schumpeter, though he praised the use of mathematics in economics, never practised it himself. Even Keynes, whose mathematical aptitude was not in doubt, avoided its use in his economics. Were these economists instinctively aware that linear mathematics was inappropriate for the representation of economic behaviour which involved primarily non-linear interactions? Would they have welcomed the application and development of a new non-linear mathematics to economics? Or is there a deeper objection to the use of mathematics? The inappropriateness of applying linear mathematics to economics was spelled out by Boulding:

> Newtonian and Cartesian numerical mathematics, which has dominated economics, is unsuitable to the more structural and topological relations of evolutionary systems ... the cardinal numbers are a property of the human imagination rather than of the real world, which consists of the topology of shapes and relative sizes rather than numbers.[13]

Elsewhere, however, Boulding refers to the *general* limitations of the use of mathematics in the social sciences. Because it is

> extraordinarily deficient in verbs beyond 'equals', 'greater than' or 'less than', mathematics has introduced a kind of linguistic bias that has prevented the study of complex relationships ... Mathematical language gains in accuracy, but at the cost of abstraction from complexity, and it is sometimes better to be inaccurate about something than to be accurate about not very much.[14]

In a similar way, the grammar of the English language makes us turn a complex set of political processes into a 'non-existent noun' called 'the United States', and leads us to think of the United States as a thing rather than as a complex of processes.

If a 'new' mathematics is to be developed to serve the needs of non-human complex adaptive systems, we should be careful that it is not applied uncritically to human systems like the market economy. It is, after all, the distinctly human element which accounts for so much of the behaviour of economic systems. This distinctly human element is manifested in entrepreneurship, in learning and in purposive behaviour in all its forms, a phenomenon which is now leading to the dominance of human cultural evolution over all other kinds of evolution. The human element is observable, too, in the formation of prices; for example, a commodity will have value as a medium of exchange only to the extent that it is believed by each agent that all other agents will agree to accept it. Again, purely subjective factors account for the periods of unwarranted optimism and pessimism which are often observed in financial markets. Whereas mathematics generally insists on entities being well-defined,[15] human beings in economic systems typically find themselves having to make decisions in circumstances which are ill-defined.

The dangers of an inappropriate type of mathematics being developed will be increased if economists continue to believe that the complex adaptive system is only a metaphor or analogy drawn from biology. It is important to emphasise that, unlike the mechanical metaphors of equilibrium analysis, the complex adaptive system is a theory (however inadequate) of how a market economy works, and not an analogy.

Mathematics has also been used in economics, not just as a language for presenting theory but as a measurement device in the search for quantitative relationships. It is still very widely believed that the hallmark of a science is its ability to predict. The capability of physicists to be able to predict within a few seconds the time of the arrival on Mars of an artificial satellite launched from Earth 11 months earlier has impressed itself on the minds of social scientists and the general public alike. Many economists have fallen into the trap of supposing that if they can imitate the methods of physics by establishing stable relationships between two or three interdependent variables, they will achieve similar successes. They do not understand that the reason for the successes of physics is not because the methods of physics are more advanced, but because the phenomena being investigated are much simpler than those of the social sciences.

In the field of biology, which is after all no less of a science than physics, there is no aspiration to prediction, and not much mathematics. The theory of evolution does not pretend to predict what forms of life will next be evolved, nor when such things will happen. To be able to make such a prediction would require a knowledge of a vast amount of dispersed information, including knowledge of events which have not yet happened. Darwin's theory of biological evolution is nevertheless falsifiable since it is capable of pattern predictions. In the case of even more complex systems like human economies, the persistent regularities which are observable have to be explained in terms of the twin principles of self-organisation and evolution, not in terms of hypothetical states of equilibrium determined by ascertainable data.

The fundamental problem with macroeconomic equilibrium theory is that the variables of its models represent aggregates or averages. It is a mistake to expect to find stable quantitative relationships among aggregates or averages in a complex adaptive system such as a market economy. While such magnitudes may be of historical interest, they are of no value in trying to explain the behaviour of the system. It would be like trying to understand the physiological functions of the human body, by measuring observed variations in the sizes of organs like the lung or the stomach.[16]

It is not therefore surprising to learn from a recent survey of the literature of macroeconomics that the results of these models are highly sensitive to the precise specification of some of their parameters.[17] As it is unlikely that these parameters can ever be estimated with precision,[18] this renders the models useless for policy purposes, which is the only justification for their existence. These considerations are unlikely to deter practitioners. What Yule called 'the lust for measurement' means that if the theoretically desirable cannot, for one reason or another, be measured, something else will be measured in its place.[19]

The explanatory power of the complexity approach

What aspects of economic behaviour in the market economy does the complexity approach help to explain? The idea of the economy as a complex adaptive system first of all focuses attention on adaptation. It shows how adaptation at the level of the individual firm and household leads to the emergence of regularities or patterns at the aggregate level, patterns such as the network of markets which performs the coordinating function between the behaviour of individuals. In this process of adaptation a key part is played by entrepreneurs. Competition is then seen as a process of entrepreneurial discovery, the outcomes of which include new products, and new processes of production and distribution.

Another regularity to emerge at the aggregate level from interactions at the individual level is the pattern of economic growth, which is a pattern of increasing specialisation, implying increasing complexity. In this process of growth, productivity depends in part on the scale of the whole economy. This implies that aggregate output is more than the sum of the output of the individual parts, which in turn suggests that there may be a contribution by society to the net social dividend.

The emergence of these and other patterns illustrate the principle of self-organisation. The other basic principle governing the behaviour of complex systems is evolution. The behaviour of individuals, including entrepreneurs, in market economies is guided by a set of formal and informal rules known collectively as institutions. For a long time, economists have taken institutions for granted, but recent events in developing countries and particularly in formerly planned economies have illustrated the importance of institutions in determining the performance of the economy. The complexity approach shows how such institutions evolve as part of the process of human cultural evolution.

The market economy is an evolutionary process whose growth is marked by increasing specialisation over time. The more specialised or complex is the economy the greater are likely to be the costs of adjustment. In the contemporary advanced economies, as competition is intensifying between countries, between companies, and between individuals in the labour market, the insecurity of employment which has always afflicted manual workers now extends to white-collar workers.

While increased labour market flexibility has produced undeniable gains in overall productivity, it has also inflicted significant social costs. These include more frequent spells of unemployment, family stress and the anxieties brought about by rapid change. In the last twenty-five years, large numbers of workers in the United States have lost health care and other job-based fringe benefits as well as a sense of job security. These changes have provoked political resistance to further rounds of trade liberalisation orchestrated by labour unions, environmentalists and other interest groups. Paradoxically, such resistance might be expected to intensify further should the growth of the US economy falter. Ways must be found to diminish the social costs of adjustment without reversing the trends towards deregulation and trade liberalisation. At the very minimum, this means understanding more about the process of adjustment in the contemporary market economy. Complexity theory focuses on adjustment, while equilibrium theory assumes all problems of adjustment away.

The complexity approach provides a framework for a unified methodological approach in the social sciences. At the present time, there is a dichotomy in economic discourse. On the one hand there is a body of pure theory, expressed largely in the form of mathematical logic, which has a life of its own. It is seldom translated into English since to do so would reveal an embarrassing gulf between the concepts in which it deals and the generally recognised features of the real world of economic activity. In short, conventional neoclassical theory, which we have called equilibrium theory, suffers from what Yeager called 'conceptual unrealism'. For this reason, it cannot usefully be deployed in addressing the common economic policy issues which daily confront business and government. There has thus grown up a second mode of discourse adopted by politicians, civil servants, City commentators and those in the media who have to deal with practical economic questions. This is chiefly characterised by the use of metaphor, with mechanistic phrases being much in evidence. Of course, equilibrium theory is itself an extended mechanical metaphor, and this is the root of the problem. This metaphor needs to be replaced by a theory which explains how the economy really works and restores such important elements of the market economy as profits, entrepreneurship, and innovation to the centre of the picture, thereby restoring some conceptual realism. The most appropriate framework currently available is the paradigm of the complex adaptive system. Amongst its advantages are the fact that it not only offers a common way of addressing both economic and business issues: it also offers a common mode of discourse with the other social sciences, and with the natural sciences as well.

Growth in the market economy is marked by qualitative change: a new product, a new technology, or a new method of distribution is introduced. This means some existing supplier suffers, so growth is not smooth but disruptive. If we add in the elements of non-linearity and chance which characterise complex systems, then periodic but unpredictable fluctuations in aggregate economic activity are the likely outcome. In the market economy, growth and fluctuations are inseparable, so the claim of some politicians to have found the formula for achieving growth without periodic recessions suggests that either they do not understand how the economy really works or else they have a new theory of the business cycle.

The complexity approach reminds us that history matters. For most participants in the market economy, the choices available today are determined in large part by the choices they made yesterday. If, at the same time, we recognise the existence of positive as well as negative feedback loops, then it becomes clear how in many business situations a company can gain a decisive advantage from being 'first-mover' in a market.

Recognition of the market economy as a self-organising system explains why it is possible to make pattern predictions about behaviour in a market economy, but it is not possible to forecast the occurrence of particular events. For example, if the price of a commodity falls, we can say with some confidence that the demand for it is likely to rise, other things being equal. But we can never say by how much, or when exactly it will happen. Likewise, we can be confident that given a significant increase in the supply of money there should follow a rise in prices, but we are unable to say by how much or when it will happen. Pattern prediction also means that it may be possible for a careful and knowledgeable observer to discern some general or particular trends in the economy, although the observer would be wise to recognise that such trends may be arrested or even reversed eventually. Thus, Schumpeter in 1942 thought he detected a long run trend away from capitalism towards socialism in both the USA and the UK. Despite the sophisticated and plausible arguments he advanced in support of his belief (see Chapter 15), this trend persisted for no more than a further decade or two before reversing itself.

In general, phenomena which emerge from an evolutionary process such as a complex adaptive system are to a large extent unpredictable. And the greater complexity of interaction which characterises human systems increases their potential for unpredictability. So the study of such systems confirms that the true objective of science is not prediction, but understanding.[20] For economists, the main reason for studying the behaviour of complex adaptive systems must be to understand the underlying interactive processes governing the behaviour of market economies.

Conclusions on the implications for economics

The congruence between the complexity approach and the Austrian approach to economics is almost complete. Together they reaffirm the restoration of the classical tradition in economic thought.

The contrast between the classical perspective and that of neoclassical equilibrium theory is most striking in their quite different concepts of the market. In equilibrium theory, a market is a central place where the aggregate supply and aggregate demand of a well-defined commodity is exchanged. In the classical view, markets emerge from the local interactions of individual trading partners each pursuing what they perceive as advantageous opportunities. During this process of interaction, these perceived opportunities evolve, either because of actual changes in circumstances, or because of changes in the perceptions of the individual agents, which is learning. In contrast to equilibrium theory, the agents' problem situation is poorly defined: they constantly search, learn and adapt to their environment.[21] As a result of adjustments at the individual level, observable patterns or regularities emerge at the level of the economy as a whole.

Instead of analysing the conditions for equilibrium for an economy for which all the data are assumed to be ascertainable, the classical approach studies how patterns and structures emerge as regularities from the process of interaction between the individual agents. The emergence, persistence and further development of such complex regularities can be explained by the two principles of self-organisation and evolution. These phenomena are not limited to economics or even the social sciences, but are perfectly general. In Hayek's words:

We understand now that *all* enduring structures above the level of the simplest atoms and up to the brain and society are the results of, and can be explained only in terms of, processes of selective evolution, and that the more complex ones maintain themselves by constant adaptation of their internal states to changes in the environment . . . These changes in structure are brought about by their elements possessing such regularities of conduct, or such capacities to follow rules, that the result of their individual actions will be to restore the order of the whole if it is disturbed by external influences.[22]

14
Implications for Business and Government

Muddling through may be the best way

Many businessmen, especially those who take the time to reflect upon what it is that they are trying to do, feel a sense of inadequacy when they compare themselves with other professional occupations. There seems to be a deficit in their understanding of the situation in which they are working; the tasks facing them seem not only interminable but intractable to analysis. They know of no alternative way of working but to proceed by intuition and hunch, or what seems no more than 'muddling through'. Whereas professions like medicine, accounting and the law have at their disposal a body of knowledge which they can rely upon to provide them with clear-cut answers in most situations, all that businessmen can fall back on are the ill-founded generalisations offered by management consultants or the equally inadequate remedies marketed by academics. These propositions are usually too vague to be helpful, and change with disconcerting frequency. Thus the thinking businessman, consciously or unconsciously, often feels intellectually inferior to his professional counterparts.

This feeling is wholly misplaced. The reason why business issues are so difficult to analyse is not because of the inadequacy of the practitioner, but because of the inherent complexity of the phenomena under analysis. The phenomena with which businessmen have to deal are not only more complex than those confronting physicists, but the data are changing much more

rapidly. The speed of light can be counted on to remain unchanged for the next 500 years. There are few such constants in
business, even from one year to another. Once the greater complexity of the phenomena being encountered by businessmen is
understood, then 'muddling through' may be seen to be a good
strategy of behaviour.

Nevertheless, businessmen are frequently exhorted by consultants and academics to adopt a more 'rational' approach to
problem-solving. The rational approach requires first identifying
general objectives, and then separately isolating the means to
achieve them. The analysis attempts to be comprehensive: every
important relevant factor is to be taken into account. Some theory
or technique is then introduced to discover the best strategy for
solving the problem. The test of a good strategy according to
this approach is that it can be shown to be the most effective
means to achieve the desired ends. Such an approach to problem-
solving may be all very well for addressing low-level problems
like the marketing of a particular product to particular segments
of the population, but once the more complex phenomena addressed at higher levels of decision-making are encountered, the
'rational-comprehensive' method turns out to be both impracti-
cal and irrelevant. In complex problems, the data requirements
of this method are impossible to attain, and the formulation of
general objectives is irrelevant.

In these more complex situations, an alternative approach is
likely to be more effective. Here, the businessman chooses between alternatives in which ends and means are closely intertwined.
He need not be concerned with any values other than incremental differences between the alternatives. The test of a good
strategy then becomes one that the executives concerned find
themselves agreeing on, without having to agree that it is the
most appropriate means to a shared objective. This procedure
Lindblom calls the Method of 'Successive Limited Comparisons',
but in business it would more commonly be recognised as 'muddling-
through'.[1] We return to the subject of strategy later in the chapter.

It is often remarked that there exists a small number of people
in business who appear to have a gift for successfully solving

complex problems by the application of hunch or intuition. This is an area of behaviour which has largely been ignored by academic researchers and others who are wedded to 'rational' (that is mechanistic) processes of thought. Recently, however, a psychologist has observed that the thought processes of an intuitive manager appear to have some parallels with the unconscious trial-and-error learning behaviour of a child.[2] Confronted for the first time with something like a Rubik's Cube or a personal computer, children use an implicit learning approach in which they play with the object, and acquire without conscious thought an understanding of the patterns that emerge. It is noticeable that children often cope with computer technology better than their parents, who usually try to find logical explanations for how things work.

Research into managerial behaviour suggests that implicit learning is at its most effective when the patterns are complex – multiple sources of information, few apparent rational links, and counter-intuitive. Those with intuitive ability are tolerant of this confusion and do not try to squeeze the information into a 'rational' process. Of course, intuitive explanations are much more difficult to communicate and are far from being infallible, so it is perhaps not surprising that the great majority of people feel more comfortable with conventional thought processes.

A somewhat similar trial-and-error approach to problem-solving in the presence of complexity is described by De Geus:

> Each company takes one step at a time. Each decision is followed by an action, and then new observations about the effect of that action, and then another step tomorrow. Before taking each new step, the company looks up and decides where to put its foot in the light of the conditions of the moment . . . Such a company knows that it is capable of only certain accomplishments at this moment in its history. These capabilities restrict the number of places where it can put its foot today. It may also have some untapped potential for future actions. Developing that potential will increase the number of places where it can step tomorrow.[3]

The absence of universally valid laws combined with the urgent need of management for instant solutions to problems has created a market for pseudo-scientific remedies which are willingly supplied. Because they cannot be universally applicable, these remedies are necessarily short-lived. Given the right circumstances, the latest fad may help some organisations solve some of their problems some of the time, but it will invariably fail to fulfil the excessive claims made for it, and will eventually be replaced to reappear in a new guise some years later.[4] With the exception of those who have specialist expertise in specific fields of technology or industry, the services of management consultants are usually a waste of shareholders' money. Their rhetoric is no substitute for the real thought which only management itself can combine with an understanding of its own firm's circumstances. In particular, no business should ever pay money for a forecast of its market or of the economy as a whole. It would be better and cheaper to make its own. Even if one made the simplest assumption that next year's values were likely to be the same as this year's (or, if a yes/no decision is required, toss a coin), it would still be likely to outperform an external consultant.

Business strategy

Some respectable business theorists (De Geus, Mintzberg, Parker and Stacey) have argued that recognition of the complexity of the market economy means that strategic planning in business is a waste of time, maybe even dangerous. Long-term business strategies, they say, can only be planned if each business decision has a limited number of predictable outcomes. The fatal weakness of intentional strategies is that they are relatively inflexible plans for wholly unknowable futures. Experts like these conclude that in an environment of complexity, companies should do no more than behave in a purely reactive manner to the unforeseeable threats and opportunities which periodically present themselves. There is no scope for purposive behaviour, or in other words for the company to act with what has been called 'strategic intent'.[5]

De Geus illustrates his argument by referring to the familiar metaphor of steering a ship. In a business organisation, as on a ship, there is a defined command structure with every individual assigned a specialist task. The captain (chief executive) exercises the control necessary to coordinate the efforts of the crew (the workforce), and sets the course (strategic direction) of the ship for its chosen destination. Although this is a metaphor which is popular with chief executives and company chairmen, in particular when composing the Annual Report ('changing tack', 'weathering the storm'), De Geus believes it is an inappropriate one. In real business life there are no charts and there is no destination. Looking back, the ship may leave behind it a distinctive wake indicating its past course, but ahead there is only an uncharted wilderness of ocean.[6]

Unlike the ship, knowledge of the business environment is dispersed throughout the company and to some degree tacit. When the environment is changing rapidly and unpredictably, control exercised from the top may therefore result in the wrong choices being made, perhaps with fatal consequences for the company's long-term survival. In such circumstances, there is an argument in favour of decentralisation of decision-making, with authority devolved to local business units, who are presumed to be in closer touch than the centre with day-to-day developments in all the markets in which the company is present. But, without central control, how can coordination of the efforts of the different business units be achieved?

Here is where the principle of self-organisation comes in. Observation of flocking behaviour in birds and shoaling in fish has shown that coordination of quite complex behaviour can be achieved without central or external control. Experiments using computer simulations have confirmed that coordinated group behaviour can be achieved by individual members of the group following certain simple rules.

The simulation starts with a collection of autonomous birdlike agents, known as 'boids', scattered randomly around the computer screen. When the experiment begins they will collect themselves into a flock that will fly around obstacles in a fluid and natural manner. Sometimes, when confronted by an obstacle, the flock will break into sub-groups that flow round both sides of the obstacle, rejoining on the other side as if planned.

But such group behaviour was *not* planned. It was the outcome of three simple rules which referred only to individuals and only to what each could see and do in its own immediate vicinity. These rules were as follows:

1. Each individual had to try to maintain a minimum distance from other objects in the environment, including other boids.
2. It had to try to match velocities (speed and direction) with boids in its neighbourhood.
3. It had to try to move towards the perceived centre of the mass of boids in its neighbourhood.

Nothing in the computer code embodying these rules of behaviour obliged a flock to be formed, and yet one formed every time. This is an example of what is called an 'emergent' phenomenon. Nor did anything in the code attempt to predict all the possible eventualities with which the boids would have to cope. Given the large number of boids, and the on-screen environment full of walls and obstacles, writing such a programme would have been an almost impossible task. It was also unnecessary.

It is tempting to draw conclusions for business strategy from such an experiment. For example, it would be possible for a chief executive to issue his individual business units with a simple rule such as 'try to make a profit', and then allow each to adapt to the changing business environment in its own way. This would imply a 'bottom-up' rather than a 'top-down' approach to strategy-making. Steering from the top is not only undesirable: it is unnecessary. This would appear to be the logical conclusion of the line of thinking embraced by Parker and Stacey, De Geus and Mintzberg.

But such reasoning is founded on analogies with the behaviour of biological but non-human complex systems. It seems to overlook the distinctive human capabilities to act purposefully and to try to alter our environment, capabilities not possessed by other living creatures but characteristic of business activity. Once the additional *human* level of complexity is recognised, there is scope for a more fruitful application of the principle of self-organisation. If a company wishes to develop a strategic objective, this can be determined at the centre and communicated to the business units. Implementation of the strategy can still be carried out on a decentralised basis, with local adaptation and feedback. In this way, decentralised implementation and learning can be made compatible with the central determination of strategy.

That this is not simply a hypothetical consideration is indicated by recent developments in thinking about the organisation of the modern professional service firm. Here, it is claimed, the trick is 'to guide people without managing them' or, as the senior partner of a law firm put it: 'Leadership is a matter of being clear on the objective. If you're very focused on that, you can be very flexible. A lot more people can come up with their own solutions. My job is to sit there and tweak people back towards the end result.'' We learn from the same article that according to the management writer Tom Peters the professional service firm is 'the best model for tomorrow's organisation in any industry'.

More generally, the importance of the part played by human direction in any organisation needs to be recognised in theories of business and economics. The performance of many firms appears to be determined in large part by the personalities and ideas of their current leaders. But, at the same time, all firms are constrained by the requirement to adapt to meet the changing needs of the markets in which they operate. Ultimately, the practices dictated from the top must be aligned with those which are consistent with the changing needs of the market if the firm is to survive. This proposition can be expressed in the language of complexity theory by saying that a complex adaptive system with a component of directed evolution can become part of a higher-order complex adaptive system in which the character of the human direction can itself evolve.

Organisation

The big question in business organisation to which every chief executive would like to know the answer is: 'What form of organisation will allow us to get the best out of our people?' This is the wrong question. Its underlying assumption is that people, like plant and raw materials, are resources to be managed in a rational machine-like manner, and that the 'right' formula is waiting to be found if only enough research is carried out. The idea of the organisation as a machine was perhaps encouraged by the experience of the Industrial Revolution. Workers were seen as being interchangeable cogs in the company's wheels, and managers were needed to keep the whole machine running smoothly. This philosophy became explicit in the 'scientific management' movement founded by the American engineer Frederick W. Taylor in the late nineteenth century. Although Taylorism seemed to fade away after the Second World War, it still raises its head in other guises, such as Business Process Re-engineering. Even the currently fashionable name for staff, 'Human Resources', has something mechanistic about it. Of course, organisations are not just machines with interchangeable parts. There are no universal formulae for organisational success. Different groups of people react differently to the same stimuli. Circumstances will determine whether particular organisational forms will be appropriate or not.

But if a business organisation is not a machine, what is it? The answer is that it is a human complex adaptive system. Unlike a machine, it cannot be controlled, but leadership can use its influence to set guiding principles and to create a decentralised environment in which good behaviours can emerge spontaneously. We know that human beings work better when they feel responsible for the goods and services they are supplying, when they have the authority to respond in appropriate ways to help those around them, whether these be customers, suppliers or colleagues; in short, when they feel part of a community. There is a great difference between service levels achieved through training and competency checklists and those which are a manifestation of inner values. It is this difference between an organisation where individual behaviour is tightly constrained by rules and one in which the right behaviour emerges spontaneously which may account for the notorious difficulty which is encountered in getting publicly owned organisations to work properly.

There is another implication of viewing the business organisation as a complex adaptive system rather than as a machine. If greater weight is to be given to the survival of the company in the long term than to the maximisation of profits, then it may be prudent in certain circumstances to tolerate a certain amount of inefficiency in the organisation. Specifically, it might be sensible to tolerate an apparent duplication of effort if it cannot be known in advance which activity will be successful and which will fail. This proposition is illustrated by De Geus with the parable of the rose bush. A grower who wishes to produce the finest rose in his district will prune his rose bushes severely in the spring. This will almost certainly produce a better result than if the bushes were left unpruned – if many shoots were tolerated. But the severely pruned bushes will be more vulnerable to such risks as disease, a late frost, or marauding deer. The roses from the unpruned bushes will never win prizes, but the bushes themselves will have the better chance of survival in the longer term and thus of producing roses in future years.[8]

Implications for government

Adoption of the complexity perspective means that governments should abandon any temptation to 'fine-tune' fluctuations in the market economy by active counter-cyclical monetary or fiscal policies. Instead of trying to change the aggregate level of output and employment they should instead try to create an appropriate environment for business. In other words, instead of trying to change the weather government should concentrate on trying to influence the climate. Specifically, this means that governments in the contemporary Western world should put themselves in the position of being able to borrow when a real recession threatens. This means running a surplus in the good times, in order to be able to run a deficit in the bad times. The borrowing could be spent on improvements to the infrastructure, which are nearly always needed.[9]

At present, it is currently fashionable to urge that training is the solution to unemployment. Some economists would go further, and propose that governments should stand ready to be the employer of last resort. Leaving aside the problem of precisely what are the skills in which training should be given, it remains true that creating a supply of skilled labour does not thereby create a demand for it.

Every government in the Western world is looking for economic policies which will lead to sustained growth in the economy that is stable and non-inflationary. This has led equilibrium theorists to search for an 'optimal' policy mix; that is, a cocktail of policies which will achieve the desired results. By implication this policy mix is of general applicability amongst countries, or at least amongst countries of a similar level of development. It follows that if Mrs Thatcher's policies of lower rates of taxation, deregulation, privatisation and monetarism were successful in Britain in the 1980s, as they undeniably were, the same formula should provide the same results in similar countries. Ireland is probably the country whose economy is most similar to Britain, yet during the same period it did not deregulate, did not privatise, adopted an incomes policy in place of monetarism, and still retains relatively high personal rates of taxation. Yet in the last decade it has achieved even higher growth rates than those achieved in Britain. Indeed, it has recently overtaken Britain in terms of GDP per head.

Market economies vary in their institutions from one to another and, within any given economy, institutions change over time so that the British economy of the 1990s is very different from the British economy of the early 1900s. So theories and policies must be developed for each specific case. And, because the economy is a complex adaptive system which is constantly evolving, policies which worked in the past will not necessarily work in the present.

The rational-comprehensive model which we noticed earlier in this chapter has periodically been applied to government policy-making, with unsuccessful results; for example, the Program, Planning and Budgetary System (PPBS) in the US in the 1960s. From time to time, it reappears often with the declared object of 'getting value for the taxpayer's money'. In the UK in the year 2000, the public sector is in the grip of the mirage of 'performance measurement'. It is characterised by taking a once-and-for-all, across-the-board view of government spending, or a large slice of it. In practice, policy is not made once and for all: it is made and remade endlessly. It is a process of successive adaptations to some desired objectives in which what is desired itself continues to change. As Lindblom points out, a policy-maker who proceeds through a succession of incremental changes can avoid serious lasting mistakes. In particular, past sequences of policy steps give the policy-maker some knowledge about the probable consequences of further similar steps.[10]

In the name of consistency, the rational-comprehensive approach would like to see a single solution applied across the board. But in the United States, for example, no agency of government attempts a comprehensive overview of policy on income distribution. Instead, a policy evolves in response to the pressures of a wide variety of interest groups. A process of mutual adjustment amongst groups representing workers and farmers, local and federal government agencies results in a distribution of income in which particular income problems neglected at one point in the decision processes become central at another. Lindblom concludes that 'for all the imperfections and latent dangers in this ubiquitous process of mutual adjustment, it will often accomplish an adaptation of policies to a wider range of interests than could be done by one group centrally'.[11]

15
The Future of the Market Economy

Many writers, from Marx onwards, have speculated about the future of the social and economic order known as capitalism. Capitalism may be defined as the market economy coupled with the private ownership of capital, although some would argue that one is inseparable from the other. Of course there is no such single phenomenon as capitalism.[1] Institutional and other arrangements vary in detail from one 'capitalist' country to another, and capitalism in each country has evolved over time, so that what we observe in the advanced market economy countries today is very different from what we should have observed in the same countries one hundred years ago. Despite these differences in detail, the complex adaptive systems which are the market economies manifest a recognisably common pattern which can be labelled as capitalism. It is the evolution over time of this pattern which is the subject of the present chapter.

By far the most scholarly treatment of the evolution of capitalism this century is to be found in the work of Schumpeter. In 1912 he provided an analysis of the characteristic pattern of behaviour of what might be called 'pure' capitalism, that is that which prevailed in the then advanced market economies at the beginning of the twentieth century.[2] Thirty years later, he provided a detailed account of the process by which he believed the typical market economy was evolving from pure capitalism through an intermediate stage of controlled or 'fettered' capitalism (what we should nowadays call the mixed economy) into the quite different pattern of organisation called socialism.[3]

The first edition of *Capitalism, Socialism and Democracy* was published in 1942 in the middle of the Second World War. The central message of the book was a simple one: that capitalism, as a system of economic and political organisation, carried within itself the seeds of its own destruction. Capitalism was doomed to fail, and the symptoms of decline were already visible in the economic and political systems not only of the United Kingdom but also of the United States. The successor state to capitalism would be some form of socialism.

This message did not arouse a great deal of interest at the time. Leading professional economists were still dazzled by the bright intellectual lights of *The General Theory*. Schumpeter found that not only his two-volume 1939 work *Business Cycles*, on which he had lavished so much scholarly effort, was apparently rendered redundant by the *The General Theory*, the same book also overshadowed his *Capitalism, Socialism and Democracy*. *The General Theory* appeared to offer an operational plan for saving capitalism, and with it democracy, while the Schumpeterian prediction had undertones of historical inevitability which were perhaps uncongenial to the Anglo-American mind.

As Haberler has pointed out,[4] Schumpeter must have gained some satisfaction from Keynes' posthumously published article[5] in which he recommended that 'the classical medicine' be allowed to work; that is, that the recession be allowed to run its course. Had he been alive today, Schumpeter would have been able to take even more satisfaction from the apparent unwillingness of the governments of the advanced countries to adopt Keynesian policies to cope with the recession of 1990–2, which in the United Kingdom was the longest recession since the Great Depression. Schumpeter would have felt that this fell into the category of 'a remedial' recession, that is, one that was necessary in order to compensate for previous errors and to provide a basis for future sustainable developments. In the words of the then British prime minister, John Major, the pains of a recession are 'a price well worth paying'.

With the general recognition of the inappropriateness of Keynesian remedies for the problems of the mixed economy, or fettered capitalism as Schumpeter would have called it, there has been a revival of scholarly interest in Schumpeter's work.[6] The essence of Schumpeter's vision of the breakdown of capitalism that is set out at length in *Capitalism, Socialism and Democracy* can be conveyed in one sentence from the book:

> The thesis I shall endeavour to establish is that the actual and prospective performance of the capitalist system is such as to negative the idea of its breaking down under the weight of economic failure, but that its very success undermines the social institutions which protect it, and 'inevitably' creates conditions in which it will not be able to live, and which strongly point to socialism as the heir apparent.[7]

Fourteen years earlier Schumpeter had put forward the same proposition:

> Capitalism, whilst economically stable and even gaining in stability, creates, by rationalising the human mind, a mentality and style of life incompatible with its own fundamental conditions, motives and social institutions, and will be changed, although not by economic necessity and probably even at some sacrifice of economic welfare, into an order of things which it will be merely a matter of taste and terminology to call socialism or not.[8]

It is significant that the author places the word 'inevitably' in the first passage within quotation marks; Schumpeter was much too wise to accept the doctrine of historical inevitability. This feature distinguishes his work from that of Marx. Schumpeter was careful to emphasise that his predictions amounted to no more than pointing out historical tendencies that could be arrested or even reversed. In general, he was sceptical of the ability of economists (or anyone else, for that matter) to predict the future.

From the two passages just quoted it is clear that Schumpeter followed Marx not only in believing that capitalism will eventually be replaced by socialism but also that this change would be brought about by factors endogenous to the capitalist system itself. But whereas, in Marx's case, capitalism would disappear in an abrupt and violent collision of economic and political forces,

Schumpeter had a very different vision. There would be a steady and gradual decomposition of pure capitalism, 'a quenching of capitalist attitudes and institutions'. According to Schumpeter, the most important factor in the decline of capitalism is the decline in the motivation of the entrepreneur, the hero of the early stage of pure capitalism. The modern business corporation, itself a product of the capitalist process, 'socialises the bourgeois mind'. The modern corporation executive acquires something of the psychology of a salaried worker employed in a bureaucratic organisation.

Meanwhile, the increasing size of firms undermines the concept of property or of freedom of contract. The figure of the working proprietor of a firm, characteristic of pure capitalism, has largely disappeared from the contemporary mixed economy. The contract of employment, which once involved a personal link between two individuals, has now become stereotyped, bureaucratic, and impersonal. In Schumpeter's words, 'the capitalist process takes the life out of the idea of property'.

An increasing concentration in the size distribution of firms also means that the political attitudes of society are profoundly changed by the elimination of a host of small and medium-sized firms. Schumpeter anticipated that the mixed economy, or 'operating the capitalist economy in the labour interest', as he put it, could not be sustained in the long run. It had the disciplines of *neither* the competitive forces prevailing in the traditional capitalist market economy *nor* the sanctions of a Soviet-style planned economy. Competition within a pure capitalist market economy eliminates all activities that no longer have an economically useful function. Such redundant activities are eliminated by decree in a planned economy (if and when they are recognised). But in a mixed economy, it is frequently possible for those who have a vested interest in uneconomic activities to use the democratic political process to protect such activities from market forces and thus to perpetuate them.

In the absence of effective incentives or sanctions, the mixed economy would therefore have to rely on self-denial or at least self-restraint on the part of those taking part in the economic process. Individuals would have to act in the interest of the economy as a whole rather than in their own material self-interest.

It would therefore appear unlikely that the mixed economy could survive for long as a viable form of economic organisation. With an increasing proportion of uneconomic activities, the pressures for change would become irresistible. Thus Schumpeter anticipated that there would be a gradual transition of the mixed economy toward a state of greater government control and regulation which he would call socialism.

Many of those who read *Capitalism, Socialism and Democracy* when it first appeared mistook Schumpeter for a socialist. Haberler says that this mistake was made even by the editor of the German edition of the book.[9] Other defenders of capitalism criticised the author's attitude as defeatist. In the preface to the second edition of his book, Schumpeter replied: 'I deny entirely that this term [defeatism] is applicable to a piece of analysis ... the report that a certain ship [capitalism] is sinking is not defeatist. Only the spirit in which this report is received can be described as defeatist. The crew can sit down and drink. But it can also rush to the pumps.'[10] Elsewhere in the book Schumpeter is discussing industries in the United Kingdom that could be taken into public ownership without serious loss of efficiency. At the bottom of the list is agriculture. Schumpeter writes:

> If he [the socialist] insists also on nationalising land – leaving, I suppose, the farmer's status as it is – that is, transferring to the state all that remains of ground rents and royalties, I have no objection to make as an economist.

To this statement, however, he adds the following footnote:

> This is no place for airing personal preferences. Nevertheless I wish it to be understood that the above statement is made as a matter of professional duty and does not imply that I am in love with that proposal which, were I an Englishman, I should on the contrary oppose to the best of my ability.[11]

It is not difficult to find the origins of Schumpeter's vision of the capitalist economic order and its counterpart, the democratic political system, as being doomed to be succeeded by the interfering collectivist state. Well before the beginning of the twentieth century, Carl Menger, the founder of the Austrian school, had become convinced that capitalism was dying. This conclusion, says Samuelson, accounted for the scholarly sterility of the last twenty-five years of Menger's life.[12]

It is interesting to speculate to what extent Menger's views concerning the demise of capitalism were influenced by his environment. Vienna at the turn of the century was a city of extraordinary social, cultural, and political change. The election of a socialist mayor in 1897 who took into municipal ownership the tramways and electric power and gas companies of the city, and the assassination of the Empress Elizabeth in 1898, seemed to portend growing instability. These changes were mirrored in the cultural and intellectual life of the city. Among the notable creative spirits in the cultural ferment of the period were the psychologist Freud, the painter Kokoschka, the composer Schoenberg, and the dramatist Kraus.

Menger passed on this pessimistic vision to the later members of the Austrian school. Both Schumpeter and Hayek, as well as von Mises, absorbed the implicit belief that capitalism was on the brink of destruction. Each of them reacted to this message in different ways. In 1927 Mises wrote surely the most intellectually devastating assault ever mounted on socialism,[13] while Hayek spent most of the later part of his life analysing the functioning of capitalism, highlighting in particular its beneficial aspects. In 1944 he published a polemical warning about the dangers of socialism.[14] This book stressed the socialist roots of Nazism, but it was an unfashionable view and the book was received with an embarrassed silence. Schumpeter was much more detached: *Capitalism, Socialism and Democracy* represents a careful analysis of the reasons for what he believed was the forthcoming decline of the capitalist system.

Superficially, Schumpeter's analysis resembles that of Marx, and he never disguised that part of his thinking that he owed to Marx. However, Schumpeter's analysis of the demise of capitalism was based on reasons entirely different from those of Marx. In Schumpeter's vision, capitalism is destroyed by its success, not its failure. The most important element in its destruction is the spread of rationalism. Capitalism would end not with a bang but a whimper, with the transition to socialism being so gradual that it would be impossible to mark clearly the boundary of the passage from the former state to the latter.

Reasons for the decline of capitalism

Pure capitalism can be defined as an economic system characterised by the private ownership of capital, the division of labour, the allocation of resources by the market, and the minimum of government intervention. In such a system, according to Schumpeter, competition takes place principally through the medium of innovation. Innovation is something brought about by competition between entrepreneurs. The driving force of competition that causes entrepreneurs to innovate is not just the prospect of profit, but the threat of loss arising from the actual or potential innovative behaviour of rival entrepreneurs. Schumpeter is careful to emphasise that it should not be supposed that innovation takes place by employing means of production which are previously unemployed. While this is possible, it is not the essence of the phenomenon. Innovation means simply putting existing factors to work in different ways.

The fact that innovation requires the bidding away of existing resources from their employment in other firms is one reason why competition through innovation is quite disruptive. Another reason is that specific forms of capital and labour with highly specialised skills rendered redundant by rival innovations elsewhere do not easily migrate to more successful firms and industries, as assumed by neoclassical theory. The value of such capital is more often destroyed, as, for example, after the switch in demand for transport services from railways to cars and airlines. In the case of specific labour skills, competition incurs significant economic costs such as the costs of unemployment and retraining as well as that of job search. Thus the typical path through time of the capitalist economy is not a smooth one, but is marked by periodic upheavals at the level of the firm and, less frequently, the industry, and by irregular fluctuations in the economy as a whole. Schumpeter's metaphor of a 'perennial gale of creative destruction' contrasts strongly with the moving equilibrium depicted by steady-state growth theory.

In *Capitalism, Socialism and Democracy*, Schumpeter explains the process according to which, in his view, pure capitalism would gradually give way, through an intermediate stage of controlled or fettered capitalism, to an economic system wholly controlled by government, which could be called socialism. Unlike Marx, Schumpeter made control and not ownership the test of socialism. Increasing government control would take three main forms: the growing regulation of business activity, increasing levels of taxation, and widening public ownership. Three further features distinguish the process of transition: an increase in concentration of the ownership of property (with a concomitant shift in political attitudes), the withering of the entrepreneurial function, and the growth of an anti-capitalist mentality.

Under pure capitalism, the Schumpeterian entrepreneur has a number of identifying features. He is moved by the desire to found a family fortune; therefore, he is content to take a long view of the required return on investment. He may also be moved by the creative satisfaction that comes from getting things done, and by the desire to prove himself superior to his competitors. He is possessed of an aptitude, entrepreneurship, that is distributed rarely among the population, like a talent for singing. He will typically launch a new firm, controlled by himself, as the vehicle for introducing an innovation. What distinguishes the entrepreneurial function from all others is that the entrepreneur innovates.

However, with the continued evolution of capitalism, the function of innovation and with it the role of the entrepreneur are expected to change. Innovation would become a routine function, carried out by existing companies, not new firms. The individual entrepreneur would gradually become extinct since his function would increasingly be performed by committees operating inside large firms. As the function of the entrepreneur becomes redundant, his or her social status would decline: so, too, would that of the whole bourgeois class. Not only would the function of innovation become routinised in this period of fettered or controlled capitalism; so, too, would the function of invention. These changes would, in Schumpeter's view, follow from the gradual extension of what he called 'the range of the calculable' as capitalism progressed.

While Schumpeter took a detached almost fatalistic, view of

what he saw as the probable evolution of the market economy, Hayek expressed his passionate hostility to any such development.[15] Hayek's objections sprang from his theory of human cultural evolution (see Chapter 9). According to this view, in the very long period of pre-history before agriculture, humans were spontaneously organised in groups or tribes of hunter-gatherers. Survival of the individual depended on membership of a group, and survival of the group depended on the instincts of solidarity and altruism of its members. When people made the transition to larger societies, instincts were not enough. They had to be replaced by rules, mainly of a self-denying nature, such as recognition of the property rights of others, rules relating to contract, and moral rules such as honesty. Hayek emphasised that these new rules were not the result of collective rational choice, but rather emerged from a process of learning by doing. Societies which adopted such rules found themselves more successful than those which did not. Rules which were perceived to 'work' were imitated and passed on, those which did not were abandoned. Thus arose the framework of institutions which made possible the market economy and its political counterpart, representative government or liberal democracy.

Hayek regarded this social order as not just another transient phenomenon of human history, but something of an optimum, something to be cherished and worth defending because it permitted the maximum degree of individual freedom while at the same time reaping the benefits of specialisation in economic activity through a system of non-coercive and mutually advantageous exchange. He saw, however, two threats to its continued survival. The first was that humans would revert to allowing themselves to be dominated by their primitive instincts. These include a distrust or hatred of outsiders, an attitude often manifested in the behaviour of totalitarian regimes. More insidious was the possibility that humans would succumb to the temptation of believing that by reason alone they could beneficially re-design the institutions of human society. In the twentieth century, this 'fatal conceit' took the form of a belief in socialism.

Although cloaked in very different language, Hayek's scenario of a possible transition from capitalism to socialism is quite similar to Schumpeter's. People who are employed in large corporations never have the opportunity to learn from first hand experience the rules of survival which are familiar to all who are engaged in small scale businesses whether as owners or employees. As an increasing proportion of the population is employed in large corporations as well as the public sector, the 'old' rules will be forgotten and the growing influence of rationalism will undermine the successful operation of the market economy. This was the path of evolution of capitalist institutions anticipated by both Hayek and Schumpeter. What has actually happened?

Consider four principal dimensions or yardsticks by which the metamorphosis of capitalism into socialism might be measured. They are: (1) the ownership of the means of production; (2) the organisation of the economy; (3) the distribution of income; and (4) values and culture.

The ownership of the means of production

From the standpoint of pure Marxism, ownership is the most important, indeed the only, acceptable criterion of progress toward socialism. While various forms of guild socialism, cooperative forms of ownership, and even municipal socialism were not only preached but actively practised in many of the advanced countries in the late nineteenth and early twentieth centuries, the orthodox Marxist position that socialism was to be equated with state ownership of all productive assets was never seriously challenged in countries that actually professed themselves to be socialist. The single exception to this rule was Tito's Yugoslavia, a deviation which, while legitimised by references to the textual works of Marx and Lenin, owed its existence in part to the particular historical circumstances in which the communist state in Yugoslavia came into existence, and in part to the personality of Tito himself. In principle, as opposed to practice, the primacy of state ownership as the sole criterion of socialism was also the subject of the interminable debates to which socialist intellectuals in the Western countries were addicted.

Marx believed that ownership of productive capital in an advanced society would pass overnight from private ownership into the hands of the state, acting on behalf of the working class, during a violent political revolution. Schumpeter had an entirely different vision of this transition, in which the question of ownership played a lesser part.

We may note with the advantage of hindsight that Schumpeter was surely right to play down the significance of ownership. When operated at arm's length, as in the case of British Petroleum for more than forty years, some French banks and insurance companies, and some Italian industrial companies, state-owned commercial organisations have not differed significantly in behaviour and performance from their privately owned counterparts. In most cases, of course, governments and parliaments in the Western world were unable to resist the temptation to interfere in the operation of state-owned companies, with consequences that were all too apparent.

Nevertheless, it must be acknowledged that, using the orthodox Marxist criterion of private versus state ownership, the progress toward socialism that seemed to characterise most, if not all, of the advanced countries in the 35 years after the Second World War, began to be reversed in the 1980s. Privatisation has become a worldwide phenomenon, spreading from the advanced to the developing countries in much the same ways as the socialist tendency had done in the immediate postwar era. Therefore, if the proportion of productive assets owned by the state is to be the criterion of the extent to which a given country has passed from capitalism to socialism, then there can be no doubt that Marx was wrong in predicting the demise of capitalism. But while Marxists have to accept this measure, Schumpeterians do not. Schumpeter clearly attached little importance to the question of the ownership of productive assets *per se*. For him, the key issue was that of government control over the private sector of the economy.

The organisation of the economy

The centrally-planned economy is the characteristic form of economic organisation associated with classical or Marxist socialism. One can justify this statement, not in terms of textual authority or philosophical argument, but by reference to the actual practices of states that professed themselves to be socialist. Once again, Yugoslavia was the exception.

To the Marxist, therefore, a second yardstick by which an economy could be said to have progressed towards socialism was the extent to which its economic activity was centrally-planned. To Schumpeter, a movement to a planned economy, although probable, was of minor significance. Instead, in his view, capitalism would gradually be suffocated under a widening and thickening blanket of regulations and taxes, responding to the political demands, via the democratic system, of interest groups, notably the trade unions.

There was plenty of evidence to be found in the first half of the twentieth century in support of his thesis, not only in the United States and the United Kingdom but in the other leading industrialised countries as well. Even as late as the mid-1970s in the United Kingdom, it might have been argued that Schumpeter's predictions had a degree of plausibility. But in retrospect it seems that the tide began to turn in the late 1970s, and that the post-war trend to ever more controlled and taxed market economies began to be reversed with accelerating rapidity in the 1980s. The counter-revolution in favour of a less restricted market economy, led simultaneously by Mrs Thatcher in the United Kingdom and President Reagan in the United States, was a programme composed primarily of removing state-owned companies to the private sector (in the United Kingdom) and of deregulation of previously regulated markets, notably the airline industry (in the United States), while in both countries a programme of decreasing rates of personal income tax was initiated. In the case of the United Kingdom, reductions were also made in the lawfully permitted powers of trade unions.

The legislative changes enacted in these two countries has significantly altered the characteristics and behaviour of the two respective economies. It has unleashed previously restrained spirits of enterprise, resulting in discernible improvements in productivity growth. At the same time some negative results have emerged such as a growing inequality in the distribution of income.

The British and American programmes of privatisation and deregulation have been widely imitated not only throughout the rest of the industrialised countries but also in the developing countries and, most dramatically of all, in the countries of Eastern Europe and the former Soviet Union. In the latter countries, living examples of Marxist socialism, the overthrow of the political dictatorship of the Communist party has been accompanied by a programme of selling state-owned industry into private hands, and the replacement of the planned economy by the free market as the principal method of allocating national resources.

In these circumstances, can it be seriously argued that Schumpeter's predictions have any validity?

Leaving aside for one moment the question of the permanence of the Thatcher–Reagan counter-revolution in the advanced countries, it can readily be conceded that the changes in the developing countries are both desirable and unlikely to be reversed. But it is the advanced countries on which Schumpeter was focusing his attention. Despite the best intentions of Mrs Thatcher and President Reagan, the growth of regulation (which can be measured for example by the number of pages in the Federal Register) shows no signs of slowing. Indeed, in recent years it has accelerated under political pressure from environmentalists and those concerned with safety at work and consumer protection. There is no reason to believe that these lobbies will grow any less influential in the future. On the other hand, the emergence of the Internet provides a large avenue for market forces to bypass and therefore frustrate many such regulations. What of taxation? Here the case for Schumpeter is stronger.

Although rates of direct taxation have been reduced, the scope and level of indirect taxation has been increased. Governments in most advanced countries are wrestling with actual or potentially chronic budget deficits, arising from an apparently insatiable demand on the part of their populations for the free or subsidised provision of welfare services, notably pensions, but also medical care, educational services, and other social security benefits. At the same time, there is a deep reluctance to pay the taxes necessary to fund such services.

Until the First World War, as Drucker points out, no government in history could obtain more than about 5 per cent or 6 per cent of the national income of its people through taxation.[16] But the growing monetisation of the economy that came with industrialisation and the development of international money markets meant that by the First World War even the two poorest combatant countries could tax and borrow in each year more than the total annual income of their respective populations. There no longer seemed to be any limits to the amounts that government could tax, borrow or spend. And, using taxes and expenditure, the government of the modern state could, above all, redistribute personal incomes.

In his 1918 paper, 'The Crisis of the Tax State', Schumpeter was the first economist to point out the implications of these possibilities. In particular, he pointed to the danger of the demand for public expenditure outrunning the willingness or the ability of society to pay for it. This could lead to a crisis:

> If the will of the people demands higher and higher public expenditures, if more and more means are used for purposes for which private individuals have not produced them, if more and more power stands behind this will, and if finally all parts of the people are gripped by entirely new ideas about private property and forms of life – then the tax state will have run its course and society will have to depend on other motive forces for its economies than self interest.[17]

By 'the tax state', Schumpeter meant what he would later call fettered capitalism, and what we should nowadays call the mixed economy.

Of the many weaknesses that afflict the mixed economy of the present day, the insatiable demand for government spending is surely the most threatening. It is a far more serious threat than the chorus of anti-capitalist sentiment which is rising once again, having been apparently silenced during the later Thatcher years. Will this chorus of criticism survive the current prosperity? My guess is that it will. The quality of modern life in the mixed economy has deteriorated so much that it cannot be repaired with the sticking plaster of another boom in output. Whatever satisfaction that may be derived from increasing quantities of consumer goods cannot compensate for the negative features of increasing inequality, congestion, and pollution that characterise daily life in the cities of the advanced countries. Nevertheless this dissatisfaction is unlikely to be critical.

Much more likely to accelerate the demise of the capitalist system, even in its fettered form, is the probable failure of governments in democratic countries to satisfy the demands of their citizens for health, medical care, and educational services as well as social security benefits. Since the wish list is open-ended, the demand is, in principle, infinite.

It was precisely this dragon that Mrs Thatcher and President Reagan set out to slay. They failed. The statistics show that for the seven leading industrialised countries in the world, every one, with the exception of the United Kingdom and Germany, experienced an increase in the ratio of government expenditure to GDP between 1979 and 1989. By 1992, both Germany and the United Kingdom had exceeded their 1979 ratio.

Apart from privatisation and deregulation, what was mainly achieved in the 1980s in the United States, the United Kingdom and elsewhere was a reduction in income tax rates (although not the overall tax burden since, especially in the United Kingdom, the burden of indirect taxes rose). But that was the politically popular part. The consequence of rising government expenditures coupled with reductions in tax rates has been a rise in government indebtedness. Again, taking the leading seven industrialised countries as a group, from 1982 to 1998, their net public debt as a percent of nominal GDP rose from 25 per cent to 45 per cent.

It is true that many circumstances do not warrant worry about rising government debt: war, recession and viable public sector investment projects, for example. However, it is surely ominous when the debt-to-national-income ratio rises during a long period without war or recession and with continually rising productivity. Furthermore, an increase in government indebtedness can set in motion processes that might threaten the stability if not the viability of the system. In other words, the build-up of debt can become sufficiently great so as to become a limiting factor in fiscal policy.

As Drucker has pointed out: The governments in all developed countries – and in most developing ones as well – have become such heavy spenders that they cannot increase their expenditures

194 Rethinking Economic Behaviour

in a recession. But that is, of course, the time when, according to all modern economic theory, they should do so in order to create purchasing power and with it revive the economy. In every single developed country governments have reached the limits of their ability to tax and their ability to borrow. They have reached these limits during boom times, when they should, according to modern economic theory, build up sizeable surpluses. The fiscal state has spent itself into impotence.[18]

When the real interest rate rises above the economy's growth rate and a primary deficit exists, then the debt ratio must rise. As the debt-to-income ratio rises, so does the risk that the debt will be repudiated.

Green puts forward an alternative view: the real problem of the contemporary mixed economy may not be a secular rise in the debt-to-national-income ratio, but instead a 'knife-edge' disequilibrium relationship between the demand for public services and the productivity of the economy. In the 1960s and 1970s too much attention, says Green, was paid to the former, at the expense of the latter. In the 1980s, the position was reversed. Certainly, this seems a correct reading of the situation in the United Kingdom, when in the late 1980s the then Chancellor of the Exchequer was using a large budget surplus to reduce the national debt. Only five years later, the size of the projected deficit constrained the freedom of movement of the Chancellor, both in the micro and macro aspects of fiscal policy.

However, Green concludes, a secular rise in the debt-to-income ratio also poses an intergenerational problem. In the short term, it simply leads to 'scrimping' on essential public services, but in the long term it can lead to civic impoverishment and 'growing inequalities between those with the wherewithal to provide substitutes for public services and those without'. Perhaps the most important intergenerational transfer is civilisation itself. Green quotes Oliver Wendell Holmes when he says that 'taxes are what we pay for civilised society'.[19]

Politicians who have said that they would increase taxes have simply not been elected. The classic example was Walter Mondale in the 1974 US presidential election. President Bush, seeking election in 1988, promised not to increase any taxes (the famous 'read my lips – no more taxes' pledge). When he did so, he was not reelected. President Clinton in the 1992 election made promises not to increase the taxation of the middle classes. Similar promises were made by Chancellor Kohl in the Federal German elections held after reunification in 1990. In the British General Election of 1997 the Labour Party pledged not to increase income tax rates in the lifetime of the Parliament.

All these candidates must have known that it was most unlikely that they would be able to keep their promises. The fact that they made them at all reflects the pressure of expectations put on them in the competition for votes that characterises contemporary capitalist democracy. Yet they were lying, as Schumpeter anticipated.[20]

On the other side of the budget, Mrs Thatcher and President Reagan were both elected on programmes of reducing public expenditure. In this case, knowing the strength of their convictions, it may be accepted that they both did truly believe in this proposition, and both tried as hard as they could to achieve a reduction. Yet two years after Mrs Thatcher left office, the proportion of public expenditure to GDP in the United Kingdom was higher than it was when she came into office. President Reagan's legacy was an increased federal government deficit as well as increased public expenditure.

Drucker claims to have detected evidence that democracy in the United States is declining as the result of elected representatives fleecing their constituents to enrich special interest groups. He points to the steady decline in voting participation in the United States, and claims that in all the advanced countries there is a concurrent decline in public interest in the functions of government, and in policy issues. Instead, he says, voters increasingly vote on the basis of 'what's in it for me?'[21]

It may be objected that the existence of a fatal flaw in democracy as a political system does not mean that capitalism is necessarily doomed. But one does not have to be a Marxist to agree that contemporary forms of Western democracy are inextricably linked to the capitalist system. Indeed, this argument is advanced with as much vigour by proponents of the market economy as it is by those opposed to it.

The distribution of income

Despite all the rhetoric about social justice that characterises the Marxist and non-Marxist tradition in socialism, the concept of equality of income does not seem to have played a major part in Marx's own thought. Indeed, if the distribution of income in the Socialist state is to be made independent of production, and organised instead to serve the principle of 'to each according to his need: from each according to his ability', then it is evident that the resulting distribution of income would be unequal, to the extent that needs are unequally distributed throughout any given population. It will, of course, be a different distribution from the unequal distribution according to ability that would be characteristic of pure or unfettered capitalism, and different again from the post-tax distribution characteristic of the fettered capitalism of the contemporary advanced countries.

While some people in the non-Marxist socialist tradition not only preached equality of incomes but actually established short-lived experimental communities that practised absolute equality of income distribution, it is not a principle that was practised in the socialist countries themselves. Indeed, the widening gap between the living standards of the governing political class, that is, members of the Communist party, and those of the rest of society was one of the factors that finally undermined the system.

What is astonishing in retrospect is the quite open way in which, in a nominally socialist society, discrimination was practised between the elite and the rest of society. In the USSR and the other communist countries of Eastern Europe, there was privileged access for the ruling group to the enjoyment of large cars, large flats, country houses, and foreign travel. In a society where the quantity and quality of consumer goods available to the general population was notoriously inadequate, exclusive access to certain shops that were well supplied with goods not available elsewhere was reserved for members of the Communist party. Despite the fact that such behaviour was devastatingly satirised by George Orwell in *Animal Farm*, and criticised by dissident socialist intellectuals such as Milovan Djilas, the practice continued right up to the collapse of the system.

It may therefore be asked whether distribution of income can legitimately be considered as an appropriate dimension for measuring the supposed capitalist transition. It is clear that for Schumpeter, as for most other writers on this subject, pure capitalism was a state in which the distribution of income was determined by earning ability (including, of course, the earning power of owned wealth), entirely untrammelled by taxation or regulation. An essential element of the transition sketched by Schumpeter was the increasing burden of taxation laid upon the upper and middle classes to which the entrepreneurs belonged. Thus, a movement towards greater equality in the distribution of income in a society is an indicator of movement in the direction away from pure capitalism and towards socialism.

While for most of this century, rates of taxation on the upper income groups in the advanced countries have increased, after the Thatcher–Reagan counter-revolution the higher rates of tax at the upper end of the income scale in both the United States and the United Kingdom were reduced. This practice has been imitated in other advanced countries such as Sweden, long renowned for its egalitarian ethos.

As far as reducing very high marginal rates of tax on the rich are concerned, it seems to have been a case of relaxing the stranglehold on the goose's neck so that a continuing supply of golden eggs can be obtained. That the politics of envy is alive and flourishing in the advanced countries is evidenced by the promises of both President Clinton in the United States and the Labour party in the United Kingdom, in their respective 1992 election campaigns, to increase once again tax rates at the upper end of the income range. Even more remarkable has been the way in which the British monarchy has been compelled by a recent upsurge of public opinion to pay income taxes, something that it had not done for more than 50 years.

In both the United States and the United Kingdom during the 1980s, standard measures showed increases in inequality in the distribution of income. But Drucker argues that taxation has made little difference to the distribution of income. Citing Pareto, he contends that only two factors determine the distribution of income in a country: productivity growth and culture. About the

United States, Drucker claims that as long as productivity was increasing, that is, until the late 1960s or early 1970s, then income distribution became more equal. But once productivity increases began to disappear, a phenomenon that he dates from the beginning of the Vietnam War, inequality began to increase. Differences in taxation between the Nixon–Carter years on the one hand, and the Reagan years on the other, made little difference.

Drucker goes on to claim that income distribution in the United Kingdom has become less equal in the last 30 years, as productivity has stopped growing. However, the rate of growth of productivity in the United Kingdom increased in the 1980s under Mrs Thatcher.

Values and culture

The central part of Marx's vision of the transition from capitalism to socialism was the transfer of political power from the middle class to the working class. Although this has not happened suddenly and violently, as Marx had predicted, it certainly has happened, and nowhere is this revolution more clearly illustrated than in the culture and public values of the societies of the advanced countries.

In the era of pure or classical capitalism, which might be approximated in the case of the United Kingdom and the United States by the period from 1875 to 1910, the culture and values that were publicly espoused and disseminated through the media of the time were essentially those of the middle classes, as Marx never tired of pointing out. While the aristocracy may have had different values (in some respects more akin to those of the working classes), they were prudent enough to keep them to themselves, and in public paid lip service to the values of the middle classes. As Oscar Wilde observed of London's upper class society of that period: 'Hypocrisy is the tribute which Vice pays to Virtue'.

Whatever the private behaviour of individuals may have been, it is not difficult to identify some of the characteristic values that were, at least outwardly, publicly approved in the society of that period. There was a work ethic of earnest endeavour, associated with, if not parsimony, at least deferred gratification, deferred in some cases until after death. Self-denial was accompanied by a culture of duty to others, by integrity, and by public responsibility. Speech and written forms of address were modest and formal to the point of pedantry. The same was true of clothing. In the arts, there were aspirations to what was believed to be good taste, even if the performance sometimes fell short. These public values were epitomised in the public architecture, notably the town halls of the great industrial cities of England. Despite higher levels of industrial pollution than exist today, the main public places of these cities were kept clean and orderly. Such values were also reflected in the only national medium of communication of the period, the press. Later, the same values were reflected in Lord Reith's BBC.

These public values were held and propagated by the middle classes. The working classes had somewhat different values, which only found public expression in the music halls and, in the United States, in the burlesque theatres. Any wider expression was frowned upon. Working-class values of this period tended toward immediate consumption rather than saving. Speech and dress were informal, and personal behaviour tended toward generosity rather than parsimony or frugality.

One hundred years ago the counterpart of today's mass media was a small number of newspapers. Radio broadcasting only began in the 1920s, but as recently as the 1950s in the United Kingdom at least it still continued to broadcast middle-class values. In this period, the working classes were forced to listen to Radio Luxembourg or else the BBC's Light Programme. Television broadcasting did not begin in Britain until the late 1930s, but as recently as the 1950s presenters were still wearing formal evening dress on television in the evenings. Today, because of the pervasiveness of film, video and television as media of communication as well as the mass circulation tabloid newspapers, public values are much more obtrusive than they were a century ago. It is impossible to escape them, unless one is a hermit.

The values reflected in the mass media of today are the polar opposite of those of the Victorian middle classes. A commercial television station, TV AM, which won its broadcasting franchise in the United Kingdom with a declared 'mission to inform', was quickly forced to abandon its approach. It had to adopt 'a mission to entertain', as the former approach proved insufficiently attractive to viewers and thus to advertisers. In newspapers, the popular tabloids have not evolved to become more like *The Times* of 100 years ago. Quite the reverse; the self-styled 'quality' newspapers have lowered their standards to become much more 'popular' in taste and appeal, while the popular tabloids compete with each other in a circulation war of ever-lowering standards.

In the United States, opinion surveys show that the great majority of people today identify themselves as 'middle class', yet it is clear that, as in the United Kingdom, the values they appear to espouse resemble those of the Victorian working class much more closely than those of the middle class of that period.

Just as politicians drawn from the aristocracy felt obliged one hundred years ago to pay lip service to middle-class values in order to get elected, so today aspiring politicians from the middle class must pay lip service to working-class or populist values. To get the Republican nomination for president in 1988, George Bush was required to use in public populist phrases that were quite incongruous. In the United Kingdom, leaders of the Conservative party feel obliged to broadcast their belief in a classless society, a state that is as difficult to imagine as an economy without firms.

Whereas 100 years ago, the prevailing standards in values, culture, fashion and all forms of public behaviour were set by a tiny handful of middle-class or upper-middle-class people, this group has now lost all social as well as political influence. Like the former metropolitan elites in a newly independent colony, the presence of this elite group in society is tolerated, as long as they do not attempt to exert any influence over the new order.

To summarise, contemporary values and culture appear to be based on self-indulgence rather than self-denial. Hedonism has replaced the work ethic. Speech and clothing are informal. Public responsibility has been replaced with the principle of 'whatever you can get away with'. Litter and graffiti are tolerated in a growing number of Western cities. In private as well as in public life, middle-class values have been overthrown. The class war is indeed over, and, as Marx and Schumpeter predicted, the middle-class values that predominated in the heyday of unfettered capitalism have been comprehensively defeated.

It is not an accident that this should be so. Marx's vision of the circumstances in which the proletariat would overthrow the bourgeoisie would appear to be one in which the accumulation of capital was a completed process. In other words, the stage of quasi-abundance of commodities had been reached, a stationary state where it was implicit that consumer wants could be satisfied by the existing stock of capital. Technical progress as well as further capital accumulation were redundant, and the economic problem could be reduced to one of administration. The whole economy would be like one gigantic public utility, and the economic problem for a socialist society would resemble the organisation of the water supply in a country where there was an abundant rainfall.

Although advanced societies may not reach this state of quasi-abundance in the foreseeable future, we are certainly moving towards it, or perhaps approaching it asymptotically. The advanced countries are rich enough to be able to support the large fraction of their population who do not work.

In such a society of quasi-abundance – the only circumstances in which it seems that a system of classical or Marxist socialism possibly could work – there is no need for traditional middle-class values. They would become redundant. Even in our present state of semi-transition, there are some indications that the pace of effort in the more advanced countries is relaxing. In the United Kingdom and the United States, an increasing proportion of each generation of the middle class works less and plays more than its predecessors. In the advanced countries, the number of hours worked on average has shown a steadily downward trend. Williams has calculated that, when the supply of labour is measured by hours offered through a working life, the reduction in the supply in the United Kingdom between 1870 and 1980 is almost 50 per cent.[22] Even Germany and Japan are showing signs of being less workaholic than they used to be. It seems therefore entirely natural that we should be observing changing attitudes toward work, leisure, personal saving and consumption.

To sum up, then, if we are to judge any movement from pure capitalism toward socialism by the extent of state ownership –

the classic Marxist criterion – then it must be recognised that the popular view of socialism in retreat is upheld. If, however, we judge any possible transition from capitalism to socialism by the indicators that Schumpeter himself would have preferred, those representing government control of the economy rather than government ownership, then a rather different picture emerges. The evidence is not unambiguous, but present trends indicate that the Achilles heel of the mixed economy may be the demand for publicly provided goods, services and transfer payments unaccompanied by commensurate willingness to pay for them. But this does not necessarily mean that it is evolving toward socialism in most of the conventionally understood senses of that word.

Only since 1989 have people in the West realised just how great a failure conventional socialism in Eastern Europe had been. Few people, with the possible exception of Hayek and von Mises, had anticipated the full extent of the economic, political, and environmental disaster that the socialist experiment turned out to be. Certainly Schumpeter did not. It is fairly certain, therefore, that within the foreseeable future, few people will want to repeat that particular mistake. Whereas Schumpeter had anticipated a growing anti-capitalist mentality and a withering of the spirit of entrepreneurship, an opposite trend has developed in the last twenty years. Entrepreneurs are widely admired, and it is the anti-capitalist mentality which has withered, particularly in Western Europe.

But if the term *socialism* is to be disallowed for the state into which contemporary forms of capitalism may be evolving, what term should then be used?

We want to describe the state in which working class or at least populist values prevail, but where private rather than public ownership of productive assets is the norm, and the allocation of resources takes place primarily through the market rather than through central planning. This might sound like social democracy, but if self-restraint cannot be exercised by society in the matter of publicly provided health, education, pensions, and other welfare benefits, then the democratic system itself may be vulnerable.

What other tendencies can we foresee at the present time? A recent development which is likely to be influential is a change in the distribution of property ownership. One hundred years ago, in the period of pure capitalism, the distribution of property was highly skewed. A large majority of the population in the then advanced countries had no property at all, or a negligible amount. Under the brief experiments in socialism, no-one in the countries concerned had any property rights at all. Today in the advanced countries, the great majority of people own non-trivial amounts of property in various forms. These include a house, state and private pension rights, some financial savings and a stock of human capital, in the form of knowledge and skills not possessed by their ancestors.

This level of property ownership together with coincident developments in technology means that self-employment may assume an increasing importance in the future. Writing before the Industrial Revolution, Adam Smith did not anticipate the creation of the nineteenth century proletariat, a consequence of labour being hired by owner's capital. Rather, he envisaged the evolution of the market economy as being characterised by independent workers each owning or hiring units of capital, operating in other words like self-employed tradesmen do today. It may be that this pattern is about to become more common. A second consequence of the spread of property ownership is that the overwhelming majority of the population will have an incentive not to take any actions which will foreseeably put their property at risk. This means they will have incentives to adopt cultural norms which are supportive of the market economy.

There is a current tendency which may be transient, but which is observable throughout the advanced countries, for individuals to respond to the increasing homogeneity of commodities and technologies by seeking to reaffirm a group identity, whether ethnic, cultural or religious. This should be seen not as a rejection of the globalisation of the market economy, but rather as a complementary response. Recent immigrants to the United States wish to share all the practices of their indigenous neighbours, but at the same time reinforce their original cultural identity, rather than becoming 'new Americans' as the immigrants of a century earlier were anxious to do. This worries Fukuyama and others who see a risk of American society fragmenting into groups with possibly conflicting values. He points out that the promotion of 'multiculturalism' with the intention of raising the self-esteem of minorities may result in the creation of social barriers and result in conflicting norms. At the same time, economic history suggests that it is cultural minorities who have frequently provided a catalyst for the process of economic growth. Perhaps the ideal solution, therefore, is a society with a dominant culture with one or more minority cultures which are large enough to provide a stimulus.

As advanced societies continue to increase in prosperity, the more advanced ones will shortly be able to afford to introduce schemes like a negative income tax. This would provide every citizen with an unconditional minimum 'social dividend'. Looking further ahead still, it is possible to envisage a society with such an abundance of material goods and the facilities for their production, that a state resembling socialism would be feasible. Abundance is the only circumstance in which socialism is conceivable in a democratic society.

It seems as if Schumpeter's fear that capitalism may undermine itself through promoting a rational and critical spirit which would turn against its own institutions may have been exaggerated. But other fears about the tendency for the progress of the market economy to undermine these institutions have arisen. Some of these are based on the disruptive effects of new business developments. A new supermarket destroys the livelihood of some family corner stores. Competition from a faraway plant leads to the closure of a local competitor. If the local plant is large, a whole community may suffer.

Like Schumpeter, John Gray sees the market economy as a process which is undermining existing institutions.[23] In his view, however, it is not just the institutions of private property which are being undermined, but the very cohesion of Western societies. By adopting a hands-off approach to market forces, Western governments are allowing our future to be determined by new technologies whose consequences we do not fully understand, our communities to be undermined by job insecurity, our environment degraded, and international conflicts to be kindled by the increasing scarcity of natural resources. He therefore advocates some form of global governance to regulate market forces which he believes are 'out of control'. However, he does not offer any concrete examples of the regulatory measures he would like to see taken other than crude protectionism (although he denies he is a protectionist). Nor does he offer any reasons why regulatory measures which have not worked at the national level in the postwar period should be any more successful at the international level. There is no doubt that job insecurity is a real issue, but the idea of addressing it by stopping the introduction of new technologies does not seem a better idea today than it did on any of the previous occasions it has been proposed.

Other writers, like Sacks, see the market economy spilling over its natural boundaries, invading other areas of life, and destroying the moral values on which it is based. Sacks complains that in the advanced economies of today, norms of behaviour such as commitments can be broken whenever they are inconvenient (divorce, gazumping). Everything can be bought and sold (babies, human eggs and sperm), a person's worth is measured by how much they earn, and shopping has become our salvation and advertising slogans our litany.[24]

These tendencies are undeniable, but Fukuyama points out that at the same time as destroying moral values the market economy is creating new ones. Adam Smith had long ago observed that the progress of commerce promoted such virtues as punctuality, thrift and honesty. Businesses have a self-interest in trading fairly: a reputation for integrity is a commercial asset. They also have a commercial interest in training their staff not just in technical skills but in rules of social conduct which allow them to cooperate more effectively with each other. Furthermore, these rules are becoming more prevalent as business organisations move to less hierarchical structures where self-management, self-organisation and networking are increasingly important, activities which are closely related to the discovery and communication of knowledge.[25]

So the market economy as it evolves both destroys and creates the rules which govern it. What is the net effect? It is meaningless to attempt to aggregate such things as institutions, which is why the concept of 'social capital' is not a very helpful one. It is a relic of the additive, mechanistic way of looking at economic activity. In attempting an assessment of the future of the market economy, it is also misleading to think in terms of unidirectional causality, such as the assertion that developments in technology drive economic developments which in turn drive political developments, and that societies everywhere are evolving to a single end, namely the market economy coupled with liberal democracy.

It is more helpful to think of society as a network of complex adaptive systems. Some of these are nested or clustered around economic activities, others around political and social activities. All are interdependent and linked to common institutions and rules with which they interact, and which themselves evolve through time. In this perspective it may be unwise to make predictions of the future evolution of the system, for at least two reasons. First, while contemporary patterns and trends can certainly be identified, their continuation in the future will depend upon events, including historical accidents, which have not yet happened, and which in a non-mechanistic universe cannot be predicted, even probabilistically.

Second, as we are dealing with *human* complex adaptive systems, future outcomes will depend in large part on human choices, for better or for worse. Therefore, instead of asking what will happen, we should ask rather what should our choices be if we are to achieve our desired objectives. Of course, the idea that democratic societies can make sensible choices through rational collective decision-making is precisely the 'conceit' against which Hayek warned. Nevertheless, there is every indication that they will continue to attempt to do so. Furthermore, experience suggests that democratic societies will go through similar processes of trial and error and learning by doing that individuals do. Hayek and Mises argued in the 1930s and 1940s with intellectual brilliance and great passion that a planned socialist economy was not feasible. Furthermore, they warned that a market economy which was restricted with high rates of taxation, tight regulation and extensive public ownership would be likely to function much less satisfactorily than one that was not burdened to such an extent. These warnings were completely ignored in Western Europe after the War. Societies discovered their truth only through painful experience.

16
Conclusions

This book has introduced a new paradigm for understanding how a contemporary advanced market economy works. The economy has been presented as an evolving complex system of interactions amongst human beings, where institutions and entrepreneurial behaviour play a central part, and there are increasing returns to scale at an aggregate level.

The contrast with the prevailing dominant paradigm of neoclassical equilibrium theory could hardly be more stark. In that model, agents behave like robots rather than human beings. All are perfectly informed about all possible opportunities, present and future, there are no direct interactions with each other, and all are assumed to behave as if they could compute and execute instantaneously optimal choices. This is known as 'rational behaviour'. In fact, it means that the economy is treated as if it were a machine, with predictable outcomes controllable by government.

All attempts at forecasting in advanced market economies have confirmed that detailed outcomes are not predictable. Attempts by governments at control have been equally unsuccessful. The Soviet experiments in central economic planning brought about the eventual collapse of their economies. In the West, milder experiments in 'fine-tuning' market economies through taxes, government spending, regulation and public ownership in the early postwar period have gradually been abandoned. In the USA and the UK, the setting of monetary and fiscal policies has at last been uncoupled from the corresponding macroeconomic equilibrium theories.

Businessmen have never accepted the mechanistic theory of how the economy works. In so far as they have looked beyond the workings of their own particular firm and industry, they and the consultants and business schools who serve their needs have sometimes been tempted by biological metaphors. Competition is often represented as a Darwinian struggle for survival. But, although biological metaphors are perhaps closer to the truth than mechanical ones, they can't account for the distinctly human element in economic activity.

All observable phenomena can be classified in a hierarchy according to the degree of their complexity. The simplest phenomena, that is the systems whose behaviour can be described in the simplest terms, are those belonging to the physical world. Although most non-scientists are accustomed to thinking of physics as a 'difficult' subject, the laws of matter are in fact relatively simple. Chemical phenomena are slightly more complicated, while the behaviour of biological organisms is more complex still. But at the top of this hierarchy of complexity come the phenomena pertaining to the human mind and to the behaviour of human beings when they interact with each other in groups. A market economy is just one example of a human complex system.

All biological and human social systems can be represented as systems which are not just complex, but also adapt to their changing environment. They have thus been labelled 'complex adaptive systems'. Such systems exhibit the principles of self-organisation and evolution. Non-linear interactions between the individual agents in such systems, combined with the influence of chance, might be expected to give rise to confusion and disorder. In fact, order emerges spontaneously in the form of regularities and patterns at the level of the system as a whole.

In the case of advanced economies, the most remarkable regularity is the existence of a network of markets whose function is to coordinate the wants of the individual agents within the system. The emergence of markets is a leading example of the phenomenon of self-organisation. Other regularities include the pattern of growth through increasing specialisation, first identified by Adam Smith and later delineated in more detail by Allyn Young. There is also the pattern in time known as the business cycle, and the often related financial crises have their own pattern. Most of these regularities are qualitative rather than quantitative, but sometimes a quantitative pattern can be detected as in the spatial location of economic activity in the United States and in the production relationships revealed in the input–output tables of developed countries.

Complex adaptive systems in general evolve, and human complex adaptive systems like market economies may be said to co-evolve with their environment. This environment includes of course physical resources, but contemporary students of economic history appear to agree that the most important elements influencing the development of market economies have been institutions. By 'institutions' is meant the framework of rules, both formal laws and informal norms of behaviour, within which market economies operate. In the short term, these rules may be relatively constant, but in the longer term they evolve in response to changes in the market economy, while the market economy in turn evolves in response to changes in its institutional environment.

Since we are not dealing with physical or biological but human complex systems, it is essential to recognise that human beings can alter their institutions in ways that neither physical nor biological agents can. From the point of view of understanding those economies where such institutions have not yet developed adequately, it is important to understand just how such an evolution took place in the more advanced market economies of the present day.

Hayek has suggested that the institutions of human society have been the outcome of one of three processes. Some rules of behaviour, such as those reflecting a distrust of strangers, are the product of instinct, and were learned in the primitive struggles of early human history where people joined groups to improve their chances of survival. Others are the result of rational collective choice. But the most important and influential rules fall into a third category, those that came about by a gradual process of cultural evolution. Any practice introduced by an individual would be adopted if it worked to the advantage of the group as a whole. Thus such fundamental institutions of society as language, the law, markets, even government itself came into existence, in Ferguson's memorable phrase, 'as the result of human action, but not the execution of any human design'.

Hayek was anxious to stress this point because he was concerned that the limits of human knowledge of social and economic interactions were not sufficiently well understood. He was afraid that 'rational' collective choices would be made in democratic societies in ignorance of the underlying behavioural relationships, with the result that outcomes would be surprising and disappointing. Apparently attractive rules would be adopted to correct perceived defects in the economy and in society which would have unintended consequences, perhaps even the opposite to those intended.

In retrospect this view may be unduly pessimistic. Societies, like individuals, may be capable of learning from their mistakes. This is at least one interpretation of the history of economic policy throughout the Western world in the postwar period. Initial attempts at government control having produced unfavourable outcomes, most Western countries have now dismantled these controls. In general, it is the adaptability of their institutions which chiefly distinguishes the performance of successful market economies from those which have been unsuccessful. Likewise, there is no reason to suppose that democratic societies should not be able to make conscious choices, both individually and collectively, to respond to problems of the physical environment. These choices will be in no sense optimal: in a complex system, such a term is meaningless. But they should be capable of adaptation and even improvement in a trial-and-error fashion. In any case, given the now overwhelming importance of human technology, and in particular its dominating influence over the future of biological evolution, 'rational' collective decision-making is likely to become an even more important influence, for better or worse, in the future of human societies.

Now that socialism is off the agenda, at least until a stage of semi-abundance is reached, what are the prospects for the future of the market economy as the dominant pattern of economic organisation? Scholars from Schumpeter to Gray have pointed out that its evolution is undermining many of the institutions on which it was founded. It is certainly true that the patterns of capitalism we observe in various countries are very different from those observed in the same countries one hundred years ago. But optimists like Fukuyama respond that the market economy throws up new institutions as fast as the old ones are torn down.

What are the implications of the new paradigm of complexity for economic theory? Economists should recognise that complexity theory is not an 'import' from the natural sciences. Its two fundamental principles of self-organisation and evolution can be traced to seventeenth and eighteenth-century philosophers of society. The view of the economy as a complex adaptive system is entirely congruent with the methodological approach of Menger and the Austrian School. It explains the coordinating function performed by the market economy, as well as its growth through increasing specialisation, phenomena identified but not explained by Adam Smith. It therefore may also be seen as a restoration of the classical tradition in economic thought.

Where does this leave neoclassical equilibrium theory? In its elementary form, as proposed by Marshall, it remains a powerful tool, particularly as an aid to teaching. However, as Marshall anticipated, its mathematical refinement has led students further away from, rather than towards, an understanding of how a market economy actually works.

The complex systems paradigm also provides a bridge between economics and the study of business (that is the study of the behaviour of the individual firm). These two subjects deal with the same phenomena but from different perspectives. They have long been divided by the methodology of equilibrium theory. Once this is abandoned, it will be possible to explore the common phenomena with a common framework of theory and a common language. The complexity paradigm also provides a common methodological framework with other social sciences and therefore holds out the prospect of progress towards a unified social theory. In practical terms, this means studying how interactions between economic and political agents can effect the performance of business and government.

Adoption of the complexity approach has many other advantages. In the course of explaining how a market economy really

212 *Rethinking Economic Behaviour*

works, complexity theory shows why prediction of particular economic events is not possible.

Because it focuses on the process of adjustment to change, it is well-suited to studying some important contemporary issues, such as the unemployment and job insecurity created by the process of competition. It explains why the more complex an economic system is, the more costly is adjustment. At the same time, it draws attention to possible solutions such as self-employment or having the government act as an employer of last resort. The episodic nature of growth in complex systems and its emphasis on the adaptation of social institutions appears to fit the facts of historical experience.

Notes

1 Introduction

1 P.W. Anderson, K.J. Arrow and D. Pines (eds), *The Economy as an Evolving Complex System*, New York: Addison-Wesley, 1988. W.B. Arthur, S.N. Durlauf and D.A. Lane (eds), *The Economy as an Evolving Complex System II*, Reading, Mass.: Perseus Books, 1997. M. Gell-Mann, *The Quark and the Jaguar*, London: Little, Brown & Co., 1994. P. Ormerod, *Butterfly Economics*, London: Faber & Faber, 1998.

2 The Economy as a Machine

1 N. Kaldor, 'The Irrelevance of Equilibrium Economics', *Economic Journal*, vol. 82, 1972, pp. 1237–55.
2 A. Marshall, *Principles of Economics*, 8th edn, London: Macmillan, 1920, p. xv.
3 Marshall, *op. cit.*, p. 461.
4 Schumpeter continued: 'The habit of applying results of this character to the solution of practical problems we shall call "The Ricardian Vice"', J. A. Schumpeter, *History of Economic Analysis*, New York: Oxford University Press, 1954, pp 472–3.
5 F.A. Hayek, *Law, Legislation and Liberty*, Vol. 3, London: Routledge & Kegan Paul, 1979, p. 68.
6 Kaldor, *op. cit.*, p. 1238.
7 Kaldor, *op. cit.*, p. 1239.
8 R.R. Nelson, 'Recent Evolutionary Theorising About Economic Change', *Journal of Economic Literature*, vol. 33, no. 1, March 1995.
9 F.A. Hayek, 'Scientism and the Study of Society', Part I, *Economica*, August 1942; Part II *Economica*, February 1943.

3 The Pretence of Knowledge

1 W.A. Sherden, *The Fortune Sellers*, New York: John Wiley, 1998, chap. 3.
2 OECD, *Economic Outlook*, vol. 53, June 1993, pp. 49–54.
3 Sherden, *op. cit.*, chap. 4.
4 Sherden, *op. cit.*, p. 116.
5 C.P. Kindleberger, *Manias, Panics and Crashes*, 3rd edn, New York: John Wiley, 1996.
6 A. Smith, *The Wealth of Nations* (1776), New York: Modern Library, 1937, p. 107.

7   Thomas F. Woodlock, quoted in *Financial Times*, 11 August 1999, Business Classics p. 35.
8   W.B. Arthur, J. H. Holland, B. Le Baron, R. Palmer and P. Tayler, 'Asset Pricing Under Endogenous Expectations in an Artificial Stock Market', in Arthur, Durlauf and Lane (1997), pp. 15–44.
9   *Ibid.*
10 Arthur *et al. op. cit.*, p. 18.
11 A. Smith, *Essays on Philosophical Subjects* (ed. W. Wightman, J. Bryce and I. Ross), Oxford: The Clarendon Press, 1980, p. 322.
12 M. Abramovitz, 'The Economics of Growth', in B.F. Haley (ed.), *A Survey of Contemporary Economics*, Homewood: Richard Irwin, 1952.
13 Abramovitz, *op. cit.*, p. 133.
14 F. H. Hahn, and R.C.O. Matthews, 'The Theory of Economic Growth: A Survey', in AEA *Surveys of Economic Theory*, vol. 2, New York: Macmillan, 1965, p. 1.
15 P.M. Romer, 'Endogenous Technological Change', *Journal of Political Economy* vol. 98, 1990, pp. 71–102.
16 M. Abramovitz, *Thinking About Growth*, Cambridge: Cambridge University Press, 1989, p. 100.
17 However, its brief existence was not without its amusing side. When the exotic catchphrase 'post-neoclassical endogenous growth theory' first appeared, it attracted the interest of some politicians. Gordon Brown, then Shadow Chancellor of the Exchequer, unwisely invoked it in a keynote speech at the 1994 Conference of the British Labour Party in justification of a greater degree of government intervention in the economy. Clearly, the mathematics was beyond the capacity of his advisers. The theory's supporters, hitherto understandably reticent about its implications for the real world, felt obliged to write to the *Financial Times* to point out that 'Overall, the new theory . . . supports the general thrust of . . . economic policies such as privatisation and deregulation', Robert Barro, *Financial Times*, 1 November 1994.
18 The former British Chancellor of the Exchequer Nigel Lawson used to refer unkindly to econometrics as 'voodoo economics' and 'witchdoctory'. This is a little unfair on witchdoctors, who may sometimes get their predictions right, especially if they time the beginning of their ceremonial rain-dances to coincide with the gathering of rainclouds on the horizon. For econometrics, a better analogy is with alchemy. No matter how often the experiments are repeated, the desired outcomes can never be attained as the underlying assumptions are false.

4 Economics and Politics

1 The real problem with inflation targeting, however, is that interest rate changes have a bigger and more immediate impact on output than they do on inflation. Common sense suggests that if it is necessary to have a monetary policy at all, it would be preferable to have one which minimised the variability of output rather than of prices. While it may be desirable that there should be no general upward trend in prices in the long run, there seems no reason why prices should not be allowed to fluctuate from year to year, if that would contribute to a greater stability of output and employment.

2 Terence Burns, 'The New Consensus on Macroeconomic Policy', *Monthly Economic Review*, Lombard Street Research, London, June 1999.

3 *Financial Times*, 15 April 1999.

4 Tim Congdon, 'Fashion and Continuity in British Fiscal Policy', *Economic Affairs*, vol. 19 no. 1, March 1999, pp. 18–23.

5 D. Brash, 'New Zealand's Remarkable Reforms', Occasional Paper no. 100, London: The Institute of Economic Affairs, p. 199.

5 Economics and Business

1 To make matters worse, they are paid higher salaries than their counterparts in other academic disciplines.

2 In this respect, there is a parallel between this process and the 'Ricardian Vice' practised by equilibrium economists (see p. 17).

3 *Financial Times*, 16 November 1995.

4 J. Kay, *The Business of Economics*, Oxford: Oxford University Press, 1996.

5 O.E. Williamson, *The Economic Institutions of Capitalism*, New York and London, 1985.

6 J. Kay, 'In Out of the Cold', *Financial Times*, 11 October 1996.

7 S.R.H. Jones, 'Transactions Costs and the Theory of the Firm', in Casson and Rose (1997), p. 24.

8 M. Casson and M. Rose, *Institutions and the Evolution of Modern Business*, Ilford: Frank Cass, 1997.

9 P. Gerowski, 'The Growth of Firms in Theory and Practice', Discussion paper no. 2092, Centre for Economic Policy and Research, London, March 1999, p. 5.

6 From Mechanical to Biological Analogies

1 P. Ormerod, *Butterfly Economics*, London: Faber & Faber, 1998, chap. 1.

2 A. Marshall, *Principles of Economics*, 8th edn, London: Macmillan, 1920, p. xiv.

3 Marshall, *op. cit.*, p. 461.

4   G.M. Hodgson, *Economics and Evolution*, Cambridge: Polity Press, 1993, p. 32.
5   As soon as we move from linear to non-linear mathematical systems, we effectively lose the capacity to predict.
6   *Financial Times*, 5 September 1997.
7   R. Dawkins, *The Selfish Gene*, New York: Oxford University Press, 1976, p. 204.
8   E. Penrose, 'Biological Analogies in the Theory of the Firm', *American Economic Review*, vol. XLII, 1952, pp. 804–19.
9   A. De Geus, *The Living Company*, Boston: Harvard Business School Press, 1997.
10 M. Gell-Mann, *The Quark and the Jaguar*, London: Little, Brown & Co., 1994.

7   The Economy as a Human Complex Adaptive System

1   F.A. Hayek, *Law, Legislation and Liberty*, vol. 1, London: Routledge & Kegan Paul, 1973, p. 15.
2   See, for example, M. Gell-Mann, *The Quark and the Jaguar*, London: Little, Brown & Co., 1994, pp. 27–8.
3   'Wherever we look, we discover evolutionary processes leading to diversification and increasing complexity', G. Nicolis and I. Prigogine, *Self-Organisation in Non-Equilibrium Systems*, New York: John Wiley, 1977.
4   S.A. Kauffman, *The Origins of Order*, New York: Oxford University Press, 1993.
5   Gell-Mann, *op. cit.*, p. 228.
6   J.H. Holland, 'The Global Economy as an Adaptive Process', in P. Anderson, K. Arrow and D. Pines, *The Economy as an Evolving Complex System*, Reading, Mass.: Addison-Wesley, 1988.
7   F.A. Hayek, *Law, Legislation and Liberty*, vol. 3, London: Routledge & Kegan Paul, 1979, p. 158.
8   Hayek (1979), *op. cit.*, p. 158.
9   M. Friedman and A. Schwartz, *A Monetary History of the United States 1867–1960*, Princeton: Princeton University Press, 1963.
10 S. Brittan, 'Causation in Economic Affairs', Occasional Paper no. 52, The David Hume Institute, Edinburgh, 1998.
11 Gell-Mann, *op. cit.*
12 W.J. Baumol and J. Benhabib, 'Chaos: Significance, Mechanism and Economic Applications', *Journal of Economic Perspectives*, vol. 3, no. 1, 1989.
13 D. Parker and R. Stacey, 'Chaos, Management and Economics', Hobart Paper no. 125, The Institute of Economic Affairs, London, 1994.
14 D.C. North, 'Some Fundamental Puzzles in Economic History', in W.B. Arthur, S.N. Durlauf and D.A. Lane (eds), *The Economy as an Evolving Complex System II*, Reading, Mass.: Perseus Books, 1997, p. 223.

15 J.H. Holland, *Adaptation in Natural and Artificial Systems*, Ann Arbor: University of Michigan Press, 1975.
16 Gell-Mann, *op. cit.*
17 Quoted in M. Waldrop, *Complexity*, p. 147.
18 Hayek (1979), *op. cit.*, p. 68.
19 The notion of path-dependency can be expressed in mathematical terms by saying that if a dynamic non-linear system has more than one stable critical point, the particular equilibrium toward which the system will move will depend upon the (out-of-equilibrium) path taken and upon the initial conditions.
20 North, *op. cit*, pp. 228–9; Hayek (1979), *op. cit.*, pp. 153–77.
21 North, *op. cit.*, p. 235.
22 P. Gerowski, 'The Growth of Firms in Theory and Practice', Discussion paper no. 2092, Centre for Economic Policy Research, London, March 1999.
23 Gerowski, *op. cit.*, p. 6.
24 *Ibid.*
25 Holland, *op. cit.*, pp. 184–5.
26 Kauffman, *op. cit.*
27 Gell-Mann, *op. cit.*, p. 100.
28 'Nations stumble upon establishments, which are indeed the result of human action, but not the execution of any human design', Adam Ferguson, *An Essay on the History of Civil Society*, London, 1767, p. 187.
29 For more detail, see Arthur, Durlauf and Lane, *op. cit.*, pp. 3 and 4.
30 B.J. Loasby, 'Market Institutions and Economic Evolution', Paper delivered at the Vienna Conference of the International Joseph A. Schumpeter Society, June 1998.

8 The Coordination of Economic Activity

1 A. Smith, *Wealth of Nations*, New York: Random House, 1937, Book 1, chapter 2, p. 26.
2 Smith, *op. cit.*, Book 4, chapter 2, p. 423.
3 K.I. Vaughn, 'The Invisible Hand' in J. Eatwell, M. Milgate and P. Newman (eds), *The New Palgrave Dictionary of Economics*, London: Macmillan, 1989.
4 Hayek (1979), p. 74.
5 The following paragraphs draw heavily on I.M. Kirzner, 'How Markets Work', Hobart Paper no. 133, The Institute of Economic Affairs, London, 1997.
6 As a consequence of the General Theory of the Second-Best.
7 'Hardly an author can be found, not even Keynes himself, who is so much the exact antipode of Milton Friedman in every part of the economist's theoretical vision as Carl Menger', Erich Streissler in J.R. Hicks and W. Weber (eds), *Carl Menger and the Austrian School of Economics*, London: Macmillan, 1973, p. 165.

8   F.A. Hayek, *Law, Legislation and Liberty*, vol. 2, London: Routledge & Kegan Paul, 1976, p. 117.
9   In the words of Hayek: '... each is made by the visible gain to himself to serve needs which to him are invisible, and in order to do so to avail himself of to him unknown particular circumstances which put him in the position to satisfy these needs at as small a cost as possible...', Hayek, 1976, p. 116.
10   Hayek (1976), p. 125.
11   *Ibid.*, p. 117.
12   Loasby, *op. cit.*
13   Hayek (1979), p. 77.
14   Hayek (1973), p. 49.
15   N. J. Vriend, 'Was Hayek an Ace?', Paper no. 403, Department of Economics, Queen Mary College, University of London, May 1999. There are further problems including a hypothetical Auctioneer.
16   L.Tesfatsion, 'Agent-based Computational Economics' (*http://www.iastate.edu/tesfatsi/ace.htm*).

9   The Evolution of Economic Institutions

1   J.A. Schumpeter, *Capitalism, Socialism and Democracy*, 5th edn, London: George Allen & Unwin, 1976, p. 82.
2   K.E. Boulding, *Ecodynamics*, Beverly Hills: Sage Publications, 1978, p. 49. Boulding called this D'Arcy Thompson's Law, after the early 20th century Scottish biologist.
3   See, for example, E. Penrose, 'Biological Analogies in the Theory of the Firm', *American Economic Review* , vol. XLII, 1952, pp. 804–19.
4   See the references given on pp. 152–3 of Hayek (1973) at note 33.
5   P.W. Anderson, K.J. Arrow and D. Pines (eds), *The Economy as an Evolving Complex System*, Reading, Mass.: Addison-Wesley, 1988. F.A. Hayek, 'The Theory of Complex Phenomena', in *Studies in Philosophy, Politics and Economics*, Chicago: University of Chicago Press, 1967, pp. 22–42. Hayek (1973), chapter 2. N. J. Vriend, 'Was Hayek an Ace?', Paper no. 403, Department of Economics, Queen Mary College, London, May 1999.
6   For a scholarly, instructive and entertaining account of evolution in both the biological and human social spheres, see Boulding *op. cit.*, 1978, and also K.E. Boulding, *Evolutionary Economics*, Beverly Hills: Sage Publications, 1981.
7   Hayek (1979), pp. 153–76.
8   Hayek, *op. cit.*, p. 161.
9   Adam Ferguson, *An Essay on the History of Civil Society*, London, 1767, p. 180.
10   Ferguson, *op. cit.*
11   Hayek (1979), p. 165.
12   Hayek, *op. cit.* At the same time, Hayek distances himself from

Schumpeter by insisting that what he calls 'the market order' was maintained more by the thousands of individuals who practised the new routines rather than by 'the occasional successful innovators whom they would imitate'.
13 Hayek (1979), p. 162.
14 Hayek (1979), p. 166.
15 F. Fukuyama, *The End of History and the Last Man*, New York: The Free Press, 1992.
16 D.C. North, 'Some Fundamental Puzzles in Economic History', in Arthur, Durlauf and Lane (1997), pp. 223–38. D.C. North, 'Understanding the Process of Economic Change', Occasional Paper no. 106, London: Institute of Economic Affairs, 1999.
17 Vriend, *op. cit.*, p. 12.
18 B.J. Loasby, 'Market Institutions and Economic Evolution', Paper delivered to a Conference of the International Joseph A. Schumpeter Society, Vienna, June 1998, pp. 11 and 12.
19 Loasby, *ibid.*
20 *Ibid.*
21 F. Fukuyama, *The Great Disruption*, London: Profile Books, 1999, pp. 143–4. New rules governing the conduct of drivers and their passengers evolved spontaneously.
22 Gell-Mann, *op. cit.*, p. 224.
23 Boulding (1978), p. 116.
24 Gell-Mann, *op. cit.*, p. 242.
25 *Ibid.*, p. 245.
26 Boulding (1978), pp. 114–15.
27 Gell-Mann, *op. cit.*, p. 241.
28 Boulding (1978), p. 115.

10 The Lessons of History

1 D.C. North, 'Understanding the Process of Economic Change', Occasional Paper no. 106, London: The Institute of Economic Affairs, 1999, p. 18.
2 Cited in D.C. North, 'Some Fundamental Puzzles in Economic History', in Arthur, Durlauf and Lane (1997), p. 232.
3 North (1997), p. 233.
4 *Ibid.*, p. 234.
5 M. Olson, *The Rise and Decline of Nations*, New Haven: Yale University Press, 1982.
6 J. Bradley, J. FitzGerald, P. Honohan and I. Kearney, 'Interpreting the Recent Irish Growth Experience', *Medium Term Review 1997–2003*, Dublin: The Economic and Social Research Institute, 1997.
7 D.S. Landes, *The Wealth and Poverty of Nations*, New York: W.W. Norton, 1998. F. Fukuyama, *The Great Disruption*, London: Profile Books, 1999.

8 North (1999), p. 24.
9 Landes, *ibid.*, p. 516.
10 *Ibid.*, p. 517.
11 *Ibid,*, p. 523.
12 *Ibid.*, p. 524.
13 *Ibid.*, p. 236.

11 Patterns in Economic Activity

1 A. Young, 'Increasing Returns and Economic Progress', *Economic Journal*, December 1928.
2 A. Smith, *Essays on Philosophical Subjects* (ed. W. Wightman, J. Bryce and I. Ross), Oxford: The Clarendon Press, 1980, p. 322.
3 W.B. Arthur, 'Increasing Returns and Path Dependence in the Economy', Ann Arbor: University of Michigan Press, 1994.
4 It is the fate of most economic advisers eventually to disappoint their political masters. It was Kondratiev's misfortune to have had as his political master Joseph Stalin. As a result, he was put to death for the unpalatability of his advice.
5 W.A. Sherden, *The Fortune Sellers*, New York: John Wiley, 1998, p. 72.
6 A. Smith, *Wealth of Nations*, 1937, pp. 703–4.
7 C.P. Kindleberger, *Manias, Panics and Crashes*, 3rd edn, New York: John Wiley, 1996.
8 *Ibid.*, p. 13.
9 *Ibid.*, p. 45.
10 P. Krugman, *The Self-Organising Economy*, Oxford: Blackwell, 1996.
11 H. Simon, 'On a Class of Skew Distribution Functions', *Biometrika*, 1955.
12 D. Simpson and J. Tsukui, 'The Fundamental Structure of Input–Output Tables', *Review of Economics and Statistics*, vol. XLVII, no. 4, 1965, pp. 434–46.
13 Most of the tables being compared date from the 1950s. The pattern may have altered slightly since then.

12 Adaptation in the Market Economy

1 J.A. Schumpeter, *Theory of Economic Development* (1912), New York: Oxford University Press, 1961.
2 L. von Mises, *Human Action*, London: William Hodge, 1949.
3 It may be inappropriate to apply the term 'learning' uncritically to adaptation in non-human biological systems as cognition may often be absent.
4 J. Rogers, *Investment Biker*, New York: Random House, 1994.
5 W. Greider, Interview in *Levy Institute Report*, vol. 7, no. 3, The Jerome Levy Economics Institute, Annandale on Hudson, 1997, pp. 9–13.

6 Will Hutton, *The State We're In*, London: Cape, 1995.
7 S. Ghoshal and C. Bartlett, *The Individual and the Corporation*, Harvard Business School Press, 1997.

13 Implications for Economics

1 This section draws heavily on W.B. Arthur, S.N. Durlauf and D.A. Lane, *The Economy as an Evolving Complex System II*, Reading, Mass.: Perseus Books, 1997, pp. 3–6.
2 Arthur *et al.*, *op. cit.*, p. 6.
3 F.A. Hayek, *Law, Legislation and Liberty*, vol. 3, London: Routledge & Kegan Paul, 1979, pp. 153–76.
4 F.A. Hayek, 'The Theory of Complex Phenomena', in *Studies in Philosophy, Politics and Economics*, Chicago: University of Chicago Press, 1967. F.A. Hayek, 'Scientism and the Study of Society', Part I, *Economica*, August 1942 ; Part II, *Economica*, February 1943.
5 See articles cited at note 33 in F.A. Hayek, *Law, Legislation and Liberty*, vol. 1, London: Routledge & Kegan Paul, 1973, pp. 152–3.
6 A. Ferguson, *A History of Civil Society*, London, 1767, p. 187. B. Mandeville, *The Fable of the Bees* (ed. F.B. Kaye) Oxford: Oxford University Press, 1924. A. Smith, *The Theory of Moral Sentiments*, London, 1759, Part 6, chapter 2.
7 P.J. Boettke, *The Elgar Companion to Austrian Economics*, Aldershot: Edward Elgar, 1994.
8 L.M. Lachmann, *Capital, Expectations and the Market Process*, Kansas City: Sheed, Andrews & McMeel, 1977.
9 A. Marshall, *Principles of Economics*, 8th edn, London: Macmillan, 1920, p. 461.
10 Arthur *et al.*, *op. cit.*, p. 4.
11 Ibid., p. 4.
12 L. Tesfatsion, 'Agent-based Computational Economics', (*http://www.iastate.edu.tesfats/ace.htm*).
13 K.E. Boulding, *Evolutionary Economics*, Beverly Hills: Sage Publishing, 1981, p. 86.
14 Boulding, *op. cit.*, p. 131.
15 But see B. Kosko, *Fuzzy Thinking*, London: HarperCollins, 1994.
16 Hayek (1979), *op. cit.*, p. 159.
17 P.R. Lane, 'The New Open Economy Macroeconomics: A Survey', Discussion Paper no. 2115, CEPR London, March, 1999.
18 L. Summers, 'The Scientific Illusion in Empirical Macroeconomics', *Scandinavian Journal of Economics*, vol. 93, 1991, pp. 129–48.
19 As Hayek observes: '. . . research techniques can be readily learnt . . . by men who understand little of the subject investigated, and their work is then often mistaken for science. But without a clear conception of the problems the state of theory raises, empirical work is usually a waste of time and resources', Hayek (1979), p. 201.

20 S. Kochugovindan and N.J. Vriend, 'Is the Study of Complex Adaptive Systems Going to Solve the Mystery of Adam Smith's "Invisible Hand"'?, *The Independent Review*, vol. 3, no. 1, Summer 1998, pp. 53–66.
21 *Ibid.*, p. 57.
22 Hayek (1979), p. 158.

14 Implications for Business and Government

1 C.E. Lindblom, 'Policy Analysis', *American Economic Review*, vol. IIL, June 1958.
2 G. Claxton, 'Knowing Without Knowing Why', *Psychologist*, May 1998.
3 A. De Geus, *The Living Company*, Boston: Harvard Business School Press, p. 155.
4 W.A. Sherden, *The Fortune Sellers*, New York: John Wiley, 1998, chapter 4.
5 G. Hamel and C.K. Prahalad, *Competing for the Future*, Boston: Harvard Business School Press, 1994.
6 De Geus, *op. cit.*, p. 154.
7 *Financial Times*, 14 May 1998.
8 De Geus, *op. cit.*, p. 142.
9 P. Drucker, *Post-Capitalist Society*, London: Butterworth-Heinemann, 1993.
10 C.E. Lindblom, 'The Science of "Muddling-Through"', *Public Administration Review*, vol. 19, 1959, pp. 79–88.
11 Lindblom, *op. cit.*, p. 86.

15 The Future of the Market Economy

1 M. Albert, *Capitalisme Contre Capitalisme*, Paris: Editions du Seuil, 1991.
2 J.A. Schumpeter, *The Theory of Economic Development* (1912), New York: Oxford University Press, 1961.
3 J.A. Schumpeter, *Capitalism, Socialism and Democracy*, 3rd edn, London: Allen & Unwin, 1950.
4 G. Haberler, 'Schumpeter's Capitalism, Socialism and Democracy after 40 Years', in A. Heertje (ed.), *Schumpeter's Vision*, New York: Praeger, 1981.
5 J.M. Keynes, 'The Balance of Payments of the United States', *Economic Journal*, vol. 56, 1946, pp. 172–87.
6 R. Swedberg, *Schumpeter: A Biography*, Princeton: Princeton University Press, 1991. R.L. Allen, *Opening Doors: The Life and Work of Joseph Schumpeter*, New York: Transaction Publications, 1991.
7 Schumpeter (1950), p. 61.
8 J.A. Schumpeter, 'The Instability of Capitalism', *Economic Journal*, vol. 38, 1928, pp. 361–86.

9 Haberler, *op. cit.*
10 Schumpeter (1950), *op. cit.*, p. xv.
11 *Ibid*, p. 231.
12 P.A. Samuelson, 'Schumpeter's Capitalism, Socialism and Democracy', in Heertje *op. cit.*
13 L. von Mises, *Socialism: An Economic and Sociological Analysis*, New Haven: Yale University Press, 1951.
14 F.A. Hayek, *The Road to Serfdom*, London: Routledge, 1944.
15 See especially F.A. Hayek, *The Fatal Conceit*, Chicago: The University of Chicago Press, 1989.
16 P. Drucker, *Post Capitalist Society*, Oxford: Butterworth-Heinemann, 1993.
17 C. Green, 'From "Tax State" to "Debt State"', *Journal of Evolutionary Economics*, vol. 3, no. 1, pp. 23–42.
18 Green, *op. cit.*
19 Green, *op. cit.*
20 According to Allen, Schumpeter's entries in his private diaries included such aphorisms as 'Democracy is government by lying', Allen, *op. cit.*
21 Drucker, *op. cit.*
22 B. Williams, 'Technology Policy and Employment', Discussion Paper, The Technical Change Centre, London, 1983.
23 J. Gray, *False Dawn*, London: Granta Books, 1998.
24 J. Sacks, 'Morals and Markets', Occasional Paper no. 108, The Institute of Economic Affairs, London, 1999.
25 F. Fukuyama, *The Great Disruption*, London: Profile Books, 1999.

Bibliography

Abramovitz, M. (1952) 'The Economics of Growth', in B.F. Haley (ed.), *A Survey of Contemporary Economics*, Homewood: Richard Irwin.

Albert, M. (1991) *Capitalisme Contre Capitalisme*, Paris: Editions du Seuil.

Alchian, A.A. (1950) 'Uncertainty, Evolution and Economic Theory', *Journal of Political Economy*, vol. 58, pp. 211–21.

Allen, R.L. (1991) *Opening Doors: The Life and Work of Joseph Schumpeter*, New York: Transaction Publications.

Anderson, P.W., Arrow, K.J. and Pines, D. (eds) (1988) *The Economy as an Evolving Complex System*, Reading, Mass.: Addison-Wesley.

Arrow, K.J. (1962) 'The Economic Implications of Learning by Doing', *Review of Economic Studies*, vol. 29, pp. 155–73.

Arthur, W.B. (1994) *Increasing Returns and Path Dependence in the Economy*, Ann Arbor: University of Michigan Press.

Arthur, W.B., Durlauf, S.N. and Lane, D.A. (eds) (1997) *The Economy as an Evolving Complex System II*, Reading, Mass.: Perseus Books.

Arthur, W.B., Holland, J.H., LeBaron, B., Palmer, R. and Tayler, P. (1997) 'Asset Pricing under Endogenous Expectations in an Artificial Stock Market', in W.B. Arthur, S.N. Durlauf and D.A. Lane, pp. 15–44.

Axelrod, R.M. (1984) *The Evolution of Co-operation*, New York: Basic Books.

Baumol, W.J. and Benhabib, J. (1989) 'Chaos: Significance, Mechanism and Economic Applications', *Journal of Economic Perspectives*, vol. 3, no. 1, pp. 77–105.

Bell, D. and Kristol, I. (eds) (1981) *The Crisis in Economic Theory*, New York: Basic Books.

Blatt, J.M. (1983) 'How Economists Misuse Mathematics', in Eichner, *op. cit.*, pp. 166–86.

Boettke, P.J. (1994) *The Elgar Companion to Austrian Economics*, Aldershot: Edward Elgar.

Boulding, K.E. (1978) *Ecodynamics*, Beverly Hills: Sage Publications.

Boulding, K.E. (1981) *Evolutionary Economics*, Beverly Hills: Sage Publications.

Boulding, K.E. (1987) 'The Epistemology of Complex Systems', *European Journal of Operational Research*, vol. 30, no. 2, pp. 100–16.

Boyd, R. and Richerson, P.J. (1985) *Culture and the Evolutionary Process*, Chicago: University of Chicago Press.

Bradley, J., FitzGerald, J., Honohan, P. and Kearney, I. (1997) 'Interpreting the Recent Irish Growth Experience', *Medium Term Review 1997–2003*, Dublin: The Economic and Social Research Institute.

Brash, D. (1996) 'New Zealand's Remarkable Reforms', Occasional Paper no. 100, London: The Institute of Economic Affairs.

225

Brittan, S. (1998) 'Causation in Economic Affairs', Occasional Paper no. 52, Edinburgh: The David Hume Institute.

Burns, T. (1999) 'The New Consensus on Macroeconomic Policy', *Monthly Economic Review*, Lombard Street Research, London, June.

Casson, M. and Rose, M. (1997) *Institutions and the Evolution of Modern Business*, Ilford: Frank Cass.

Casti, J.L. (1997) *Would-Be Worlds*, New York: John Wiley.

Clark, C.M.A. (1990) 'Adam Smith and Society as an Evolutionary Process', *Journal of Economic Issues*, vol. 24, no. 3, pp. 825–44.

Clark, N.G. and Juma, C. (1987) *Long Run Economics: An Evolutionary Approach to Economic Growth*, London: Pinter.

Claxton, G. (1998) 'Knowing Without Knowing Why', *Psychologist*, May.

Congdon, T. (1999) 'Fashion and Continuity in British Fiscal Policy', *Economic Affairs*, vol. 19, no. 1, pp. 18–23.

Darwin, C. (1859) *The Origin of Species By Means of Natural Selection*, London: Watts.

Dawkins, R. (1976) *The Selfish Gene*, New York: Oxford University Press.

Debreu, G. (1991) 'The Mathematisation of Economic Theory', *American Economic Review*, vol. 81, no. 1.

De Geus, A. (1998) *The Living Company*, Boston: Harvard Business School Press.

Drucker, P. (1993) *Post-Capitalist Society*, London: Butterworth-Heinemann.

Eatwell, J., Milgate, M., and Newman P. (eds) (1987) *The New Palgrave Dictionary of Economics* (4 vols), London: Macmillan.

Eichner, A.S. (ed.) (1983) *Why Economics is Not Yet A Science*, Armonk, N.Y.: Sharpe.

Ferguson, A. (1986) *An Essay on the History of Civil Society* (1767), Edinburgh: Edinburgh University Press.

Friedman, M. and Schwartz, A. (1963) *A Monetary History of the United States 1867–1960*, Princeton: Princeton University Press.

Fukuyama, F. (1992) *The End of History and the Last Man*, New York: The Free Press.

Fukuyama, F. (1999) *The Great Disruption*, London: Profile Books.

Gell-Mann, M. (1994) *The Quark and the Jaguar*, London: Little, Brown & Co.

Gerowski, P. (1999) 'The Growth of Firms in Theory and Practice', Discussion Paper no. 2092, Centre for Economic Policy Research, London, March.

Ghoshal, S. and Bartlett, C.A. (1997) *The Individual and the Corporation*, New York: Harper Business.

Goodhart, C. (1999) 'Central Bankers and Uncertainty', *Proceedings of the British Academy*, 101: Lectures and Memoirs 1998, Oxford: Oxford University Press.

Gray, J. (1998) *False Dawn*, London: Granta Books.

Green, C. 'From "Tax State" to "Debt State"', *Journal of Evolutionary Economics*, vol. 3, no. 1, pp. 23–42.

Griswold, C.L. (1999) *Adam Smith and the Virtues of Enlightenment*, Cambridge: Cambridge University Press.

Haavelmo, T. (1954) *A Study in the Theory of Economic Evolution*, Amsterdam: North Holland.

Haberler, G. (1981) 'Schumpeter's Capitalism, Socialism and Democracy After 40 Years,' in Heertje, *op. cit.*

Heertje, A. (ed.) (1981) *Schumpeter's Vision*, New York: Praeger.

Hahn, F.H. and Matthews, R.C.O. (1965) 'The Theory of Economic Growth: A Survey' in AEA *Surveys of Economic Theory*, vol. 2, New York: Macmillan.

Hamel, G. and Prahalad, C.K. (1994) *Competing for the Future*, Boston: Harvard Business School Press.

Hamowy, R. (1987) *The Scottish Enlightenment and the Theory of Spontaneous Order*, Carbondale, Ill.: Southern Illinois Press.

Hardy, A.C. (1965) *The Living Stream*, London: Collins.

Hayek, F.A. (1943) 'Scientism and the Study of Society', Part I, *Economica*, August 1942; Part II, *Economica*, February.

Hayek, F.A. (1944) *The Road To Serfdom*, London: Routledge.

Hayek, F.A. (1967) 'The Theory of Complex Phenomena', in *Studies in Philosophy, Politics and Economics*, Chicago: University of Chicago Press, pp. 22–42.

Hayek, F.A. (1973) *Law, Legislation and Liberty*, vol. 1, London: Routledge and Kegan Paul.

Hayek, F.A. (1976) *Law, Legislation and Liberty*, vol. 2, London: Routledge and Kegan Paul.

Hayek, F.A. (1979) *Law, Legislation and Liberty*, vol. 3, London: Routledge and Kegan Paul.

Hicks, J.R. and Weber, W. (eds) (1973) *Carl Menger and the Austrian School of Economics*, London: Macmillan.

Hirshleifer, J. (1977) 'Economics From A Biological Viewpoint', *Journal of Law and Economics*, vol. 20, pp. 1–52.

Hodgson, G.M. (1993) *Economics and Evolution*, Cambridge: Polity Press.

Holland, J.H. (1975) *Adaptation in Natural and Artificial Systems*, Ann Arbor: University of Michigan Press.

Holland, J.H. (1988) 'The Global Economy as an Adaptive Process', in P.W. Anderson, K.J. Arrow and D. Pines (eds), *op. cit.*

Holland, J.H. (1995) *Hidden Order: How Adaptation Builds Complexity*, Reading, Mass.: Addison-Wesley.

Hume, D. (1985) *Essays, Moral, Political and Literary* (1741) (ed. E.F. Miller), Indianapolis: Liberty Classics.

Hutton, W. (1995) *The State We're In*, London: Cape.

Jantsch, E. (1980) *The Self-Organising Universe*, Oxford: Pergamon Press.

Johnson, G. (1997) *Fire in the Mind*, London: Penguin Books.

Jones, S.R.H. (1997) 'Transactions Costs and the Theory of the Firm', in M. Casson and M. Rose, *op. cit.*

Kaldor, N. (1972) 'The Irrelevance of Equilibrium Economics', *Economic Journal*, vol. 82, pp. 1237–55.

Kauffman, S.A. (1993) *The Origins of Order*, New York: Oxford University Press.

Kay, J. (1993) *Foundations of Corporate Success*, Oxford: Oxford University Press.

Kay, J. (1996) *The Business of Economics*, Oxford: Oxford University Press.

Keynes, J.M. (1936) *The General Theory of Employment, Interest and Money*, London: Macmillan.

Keynes, J.M. (1946) 'The Balance of Payments of the United States', *Economic Journal*, vol. 56, pp. 172–87.

Kindleberger, C.P. (1996) *Manias, Panics and Crashes*, 3rd edn, New York: John Wiley.

Kirzner, I.M. (1997) 'How Markets Work', Hobart Paper no. 133, London: The Institute of Economic Affairs.

Kochugovindan, S. and Vriend, N.J. (1998) 'Is the Study of Complex Adaptive Systems Going to Solve the Mystery of Adam Smith's "Invisible Hand"'?, *The Independent Review*, vol. 3, no. 1, pp. 53–66.

Kosko, B. (1994) *Fuzzy Thinking*, London: HarperCollins.

Krugman, P. (1996) *The Self-Organising Economy*, Oxford: Blackwell.

Lachman, L.M. (1977) *Capital, Expectations and the Market Process*, Kansas City: Sheed, Andrews and McMeel.

Lakatos, I. (1970) 'Falsification and the Methodology of Scientific Research Programmes', in Lakatos, I. and Musgrave, A. (eds) *Criticism and the Growth of Knowledge*, Cambridge: Cambridge University Press.

Landes, D.S. (1998) *The Wealth and Poverty of Nations*, New York: W.W. Norton.

Lane, P.R. (1999) 'The New Open Economy Macroeconomics: A Survey', Discussion Paper no. 2115, London: Centre for Economic Policy Research.

Lindblom, C.E. (1958) 'Policy Analysis', *American Economic Review*, vol. 48.

Lindblom, C.E. (1959) 'The Science of "Muddling-Through"', *Public Administration Review*, vol. 19, pp. 79–88.

Loasby, B.J. (1976) *Choice, Complexity and Ignorance*, Cambridge: Cambridge University Press.

Loasby, B.J. (1991) *Equilibrium and Evolution*, Manchester: Manchester University Press.

Loasby, B.J. (1998) 'Market Institutions and Economic Evolution', Paper delivered at the Vienna Conference of the International Joseph A. Schumpeter Society, June.

Mandeville, B. (1924) *The Fable of the Bees* (1714) (ed. F.B. Kaye), Oxford: Clarendon Press.

Marshall, A. (1920) *Principles of Economics*, 8th edn, London: Macmillan.

Maynard Smith, J. (1982) *Evolutionary Game Theory*, Cambridge: Cambridge University Press.

Menger, C. (1963) *Problems of Economics and Sociology* (1883), Urbana: University of Illinois Press.

Menger, C. (1892) 'On the Origins of Money', *Economic Journal*, vol. 2, no. 2, pp. 239–255.

Milgrom, P. and Roberts, J. (1992) *Economics, Organisation and Management*, Englewood Cliffs, N.J.: Prentice Hall.

Mintzberg, H. (1989) *Mintzberg on Management*, New York: The Free Press.
Mintzberg, H. (1994) *The Rise and Fall of Strategic Planning*, New York: The Free Press.
Mises L. von (1951) *Socialism: An Economic and Sociological Analysis*, New Haven: Yale University Press.
Mises, L. von (1949) *Human Action*, London: William Hodge.
Monod, J. (1971) *Chance and Necessity*, New York: Knopf.
Nelson, R.R. (1995) 'Recent Evolutionary Theorising About Economic Change', *Journal of Economic Literature*, vol. 33, no. 1, pp. 48–90.
Nelson, R.R. and Winter, S.G. (1974) 'Neoclassical vs. Evolutionary Theories of Economic Growth', *Economic Journal*, vol. 84, no. 4, pp. 886–905.
Nicolis, G. and Prigogine, I. (1977) *Self-Organisation in Non-Equilibrium Systems*, New York: John Wiley.
North, D.C. (1997) 'Some Fundamental Puzzles in Economic History' in Arthur, Durlauf and Lane.
North, D.C. (1999) 'Understanding the Process of Economic Change', Occasional Paper 106, London: Institute of Economic Affairs.
Olson, M. (1982) *The Rise and Decline of Nations*, New Haven: Yale University Press.
Ormerod, P. (1998) *Butterfly Economics*, London: Faber and Faber.
Parker, D. and Stacey, R. (1994) 'Chaos, Management and Economics', Hobart Paper 125, London: The Institute of Economic Affairs.
Penrose, E.T. (1952) 'Biological Analogies in the Theory of the Firm', *American Economic Review*, vol. XLII, pp. 804–19.
Penrose, E.T. (1995) *The Theory of the Growth of the Firm*, 3rd edn, Oxford: Oxford University Press.
Peters, T.J. and Waterman, R.H. (1994) *In Search of Excellence*, Sydney: HarperCollins.
Prigogine, I. (1997) *The End of Certainty*, New York: The Free Press.
Putnam, R.D. (1993) *Making Democracy Work*, Princeton N.J.: Princeton University Press.
Radner, R. (1968) 'Competitive Equilibrium Under Uncertainty', *Econometrica*, vol. 36, no. 1, pp. 31–58.
J. Sacks (1999) 'Morals and Markets', Occasional Paper 108, London: The Institute of Economic Affairs.
Samuelson, P.A. (1981) 'Schumpeter's Capitalism, Socialism and Democracy' in Heertje.
Schumpeter, J.A. (1954) *History of Economic Analysis*, New York: Oxford University Press.
Schumpeter, J.A. (1961) *The Theory of Economic Development* (1912), New York: Oxford University Press.
Schumpeter, J.A. (1950) *Capitalism, Socialism and Democracy*, 3rd edn, London: Allen and Unwin.
Shackle, G.L.S. (1972) *Epistemics and Economics*, Cambridge: Cambridge University Press.
Sherden, W.A. (1998) *The Fortune Sellers*, New York: John Wiley.

Simon, H.A. (1955) 'On A Class of Skew Distribution Functions', *Biometrika.*
Simpson, D. and Tsukui, J. (1965) 'The Fundamental Structure of Input-Output Tables', *Review of Economics and Statistics*, vol. XLVII, no. 4, pp. 434–46.
Simpson, D. (1984) 'Joseph Schumpeter and the Austrian School of Economics', *Journal of Economic Studies*, vol. 10, no. 4, pp. 15–28.
Smith, A. (1976) *The Theory of Moral Sentiments* (1759), Oxford: Clarendon Press.
Smith, A. (1937) *The Wealth of Nations* (1776), New York: Modern Library.
Smith, A. (1980) *Essays on Philosophical Subjects* (ed. W. Wightman), Oxford: Clarendon Press.
Streissler, E.W. (ed.) (1969) *Roads To Freedom: Essays In Honour of Friedrich A von Hayek*, London: Routledge and Kegan Paul.
Sugden, A. (1989) 'Spontaneous Order', *Journal of Economic Perspectives*, vol. 3, no. 4, pp. 85–97.
Summers, L. (1991) 'The Scientific Illusion in Empirical Macroeconomics', *Scandinavian Journal of Economics*, vol. 93, pp. 129–48.
Swedberg, R. (1991) *Schumpeter: A Biography*, Princeton: Princeton University Press.
Tesfatsion, L. Agent-based Computational Economics (*http://www.iastate.edu/ tesfatsi/ace.htm*).
Vriend, N.J. (1999) 'Was Hayek an Ace?' Paper no. 403, Department of Economics, Queen Mary College, University of London, May.
Waldrop, M. (1992) *Complexity*, New York: Simon & Schuster.
Wiles, J.S., Kunkel, J. and Wilson, A.C. (1983) 'Birds, Behaviour and Anatomical Evolution', *Proceedings of the National Academy of Sciences*, vol. 80, pp. 4394–97.
Williams, B. (1983) 'Technology Policy and Employment,' Discussion Paper, The Technical Change Centre, London.
Williamson, O.E. (1985) *The Economic Institutions of Capitalism*, London: Macmillan.
Williamson, O.E. (1993) *The Nature of the Firm*, Oxford: Oxford University Press.
Wilson, E.O. (1978) *On Human Nature*, Cambridge: Harvard University Press.
Witt, U. (1985) 'Co-ordination of Individual Economic Activities As An Evolving Process of Self-Organisation', *Economie Appliquée*, vol. 37, no. 4, pp. 569–95.
Witt, U. (1995) *Individualistic Foundations of Evolutionary Economics*, Cambridge: Cambridge University Press.
Young, A.A. (1928) 'Increasing Returns and Economic Progress', *Economic Journal*, vol. 38, no. 4, pp. 527–42.

Index